COURSE OF MATHEMATICAL LOGIC

VOLUME 2

MODEL THEORY

SYNTHESE LIBRARY

MONOGRAPHS ON EPISTEMOLOGY,

LOGIC, METHODOLOGY, PHILOSOPHY OF SCIENCE,

SOCIOLOGY OF SCIENCE AND OF KNOWLEDGE,

AND ON THE MATHEMATICAL METHODS OF

SOCIAL AND BEHAVIORAL SCIENCES

Editors:

DONALD DAVIDSON, *Rockefeller University and Princeton University*

JAAKKO HINTIKKA, *Academy of Finland and Stanford University*

GABRIËL NUCHELMANS, *University of Leyden*

WESLEY C. SALMON, *University of Arizona*

VOLUME 69

ROLAND FRAÏSSÉ

COURSE OF MATHEMATICAL LOGIC

VOLUME 2

Model Theory

D. REIDEL PUBLISHING COMPANY

DORDRECHT-HOLLAND / BOSTON-U.S.A.

COURS DE LOGIQUE MATHÉMATIQUE, TOME 2
First published by Gauthier-Villars and E. Nauwelaerts, Paris, Louvain, 1967
Second revised and improved edition, Gauthier-Villars, Paris, 1972
Translated by David Louvish

Library of Congress Catalog Card Number 72–95893

ISBN-13: 978-90-277-0510-5 e-ISBN-13: 978-94-010-2097-8
DOI: 10.1007/978-94-010-2097-8

Published by D. Reidel Publishing Company,
P.O. Box 17, Dordrecht, Holland

Sold and distributed in the U.S.A., Canada, and Mexico
by D. Reidel Publishing Company, Inc.
306 Dartmouth Street, Boston,
Mass. 02116, U.S.A.

TABLE OF CONTENTS

CHAPTER 3 / ELIMINATION OF QUANTIFIERS

CHAPTER 4 / EXTENSION THEOREMS

CHAPTER 5 / THEORIES AND AXIOM SYSTEMS

CHAPTER 6 / PSEUDO-LOGICAL CLASS; INTERPRETABILITY OF THEORIES; EXPANSION OF A THEORY; AXIOMATIZABILITY

CHAPTER 7 / ULTRAPRODUCT

CHAPTER 8 / FORCING

CHAPTER 9 / ISOMORPHISMS AND EQUIVALENCES IN RELATION TO THE CALCULUS OF INFINITELY LONG FORMULAS WITH FINITE QUANTIFIERS

PREFACE

This book is addressed primarily to researchers specializing in mathematical logic. It may also be of interest to students completing a Masters Degree in mathematics and desiring to embark on research in logic, as well as to teachers at universities and high schools, mathematicians in general, or philosophers wishing to gain a more rigorous conception of deductive reasoning.

The material stems from lectures read from 1962 to 1968 at the Faculté des Sciences de Paris and since 1969 at the Universities of Provence and Paris-VI. The only prerequisites demanded of the reader are elementary combinatorial theory and set theory. We lay emphasis on the semantic aspect of logic rather than on syntax; in other words, we are concerned with the connection between formulas and the multirelations, or models, which satisfy them. In this context considerable importance attaches to the theory of relations, which yields a novel approach and algebraization of many concepts of logic.

The present two-volume edition considerably widens the scope of the original [French] one-volume edition (1967: *Relation, Formule logique, Compacité, Complétude*). The new Volume 1 (1971: *Relation et Formule logique*) reproduces the old Chapters 1, 2, 3, 4, 5 and 8, redivided as follows: Word, formula (Chapter 1), Connection (Chapter 2), Relation, operator (Chapter 3), Free formula (Chapter 4), Logical formula, denumerable-model theorem (Löwenheim-Skolem) (Chapter 5), Completeness theorem (Gödel-Herbrand) and Interpolation theorem (Craig-Lyndon) (Chapter 6), Interpretability of relations (Chapter 7).

The new Volume 2, *Model Theory*, reproduces the old Chapters 6 and 7 (renumbered as Chapters 1 and 5), and seven new chapters are added: Logical restriction theorems (Chapter 1), Compactness theorem (Gödel-Tarski), Omission theorem, Interpretability theorem (Svenonius) (Chapter 2), Elimination of quantifiers (Chapter 3), Logical extension theorems (Chapter 4), Theory, axiomatic system (Chapter 5), Pseudo-logical class, expansion, axiomatizability (Chapter 6), Ultraproducts (Chapter 7),

Forcing (Chapter 8), Logic of infinite formulas (Chapter 9). Much of the material published in this translation has been added by the author since publication of the French edition and the division of chapters is different.

ROLAND FRAÏSSÉ

TRANSLATOR'S PREFACE

The reader familiar with Professor Fraïssé's work is surely aware that his terminology differs in many particulars from that accepted by most logicians. This presents the translator with a particularly acute instance of a dilemma familiar in rendering scientific texts: should he attempt to convey the original terminology and style of the source as faithfully as possible, or rather paraphrase them, using the vocabulary most widely accepted in the target language?

In the present case I believe the first alternative to be the more appropriate. Interlingual differences in terminology are most frequently accidental. However, as I know from correspondence with Professor Fraïssé, he knowingly and purposefully employs many terms unfamiliar even to other French logicians. Indeed, he holds quite strong views on the customary terminology, considering much of it to be unfortunately chosen. For example, he criticizes the use of 'elementary' in such phrases as 'elementary class', 'elementarily equivalent', etc., preferring 'logical class', 'logically equivalent', etc., on the grounds that any concept might be called 'elementary' in contrast to more sophisticated concepts. His principle is to adopt a logical vocabulary, keeping as close as possible to tradition in order to preserve readability.

In short, Professor Fraïssé's at first sight strange terminology is not fortuitous but rather an intrinsic component of his philosophical outlook. The reader may or may not agree with the author's views, but the translator, to my mind, cannot be permitted this liberty. He therefore has no alternative but to remain as close as possible to the original French (except in isolated cases where the result would have been a glaring gallicism), though it would have been possible to replace each term by the most widespread English equivalent, as the knowledgeable reader will surely confirm.

I hope that in so doing I have succeeded to some degree in conveying the original flavor of Professor Fraïssé's stimulating work. The intelligent reader will be highly rewarded for the slight additional effort this approach entails.

DAVID LOUVISH

INTRODUCTION

This introduction is a complement to the introduction of Volume 1, refining and adding to the discussion presented there.

In regard to the history of logic, notably to its subdivision into successive steps, we feel the need for a more rigorous justification of the "chronology" we have proposed, which might conceivably be carried over to other theoretical disciplines.

(1) Characteristic of the *prehistory* of the theory is that it exhibits only the rudiments of an axiomatic system, intended to represent certain facts. Thus, Aristotelian syllogistics claimed to represent deductive reasoning; Pascal's theory of definition anticipated the concepts of noncreativity and eliminability, which were given a fully rigorous treatment much later by Leśniewski (1927; see 6.2 below).

(2) The prehistory is followed by a period which might be called *infancy*, during which one observes one or more clear-cut axiomatic systems, though these are "immature" in the sense that they do not yet pose problems of mathematical technique. Such are the propositional calculus developed by Boole and his successors, and the *Principia Mathematica* of Russell and Whitehead. Even Peano's arithmetic and Zermelo-Fraenkel set theory presented no mathematical problems when they appeared, with the natural exception of already familiar problems from the theories they represented. It is also significant that the business of axiomatization was taken up not only by certain outstanding mathematicians (Hilbert – geometry and the first attempts toward semantics) but also by philosophers, whose logical genius compensated for their lack of mathematical training (Russell). Recall that this period of infancy in logic extends approximately from 1840 to 1915.

(3) The next period is aptly designated as *coming of age*, when the theory in question begins to present difficult problems of mathematical technique. These theories are capable of further development thanks solely to the interest they arouse in mathematicians, independently of the facts and experience to which they owe their birth. This is the case in

semantic theory, beginning with the denumerable-model theorem of Löwenheim (1915) and Skolem (1920; see Volume 1, Chapter 5); in the propositional calculus, dating from Łukasiewicz (1920) and Post (1921; see Volume 1, Chapter 2). A complementary trait indicative of coming of age is the appearance of new theories, inspired by progress in the old theories but devoid of any direct link with experience, such as logical calculi with infinitely long formulas, which date back (at least) to Tarski (1958; see Chapter 9 below).

(4) The *maturity* of a theory begins when it offers solutions to problems originating in other theories, which have already reached maturity. This idea is not vacuous if we assume *a priori* that modern arithmetic, algebra and analysis are 'mature'. Logic may then claim to have reached maturity around 1965, with the publication of the Ax-Kochen theorem – the first indisputably arithmetical result achieved by indisputably logical methods.

Note that the results of Julia Robinson (1949; see Volume 1, Chapter 7 and the Appendix below), though involving an arithmetically inspired technique, are of interest only to the logician, enabling him to define the integers in the field of rational numbers. They confirm only the coming of age of logic and not its maturity. On the other hand, the consistency and independence of the axiom of choice and the generalized continuum hypothesis (Gödel, 1940, Cohen, 1963), though more important than the Ax-Kochen theorem, do not seem to imply the maturity of logic. In fact, they may either be regarded as belonging to logic, in which case they only confirm its coming of age; or they may be construed as parts of set theory – but set theory is not mature, since it has not as yet proved new and important results in arithmetic, algebra or analysis. Thus the Gödel-Cohen results do not entail the maturity of logic in the sense specified above.

The next subject for discussion is of interest for logicians, mathematicians and teachers. What are the criteria that distinguish modern mathematics from classical mathematics? Certain oft-cited criteria seem to characterize rather the mathematical "vogue" of the time than its "modernistic" aspects; indeed, they are sometimes unjustifiably viewed as factors of progress, whereas they are actually indifferent or even harmful to progress.

According to a widespread opinion, an axiom system is modern if it is *multivalent*, in other words, if it admits several nonisomorphic models; a classical axiom system, on the other hand, is univalent (admits only one model, up to isomorphism). In the usual logical calculus there are no univalent theories, except those describing a finite model, which are of little interest. Of the theories with infinite models, those approaching most closely to univalence are theories which are categorical in power, having only one model for each infinite cardinal (such as the theory of commutative groups of characteristic 2; see 5.4 below). Peano arithmetic, a very faithful representation of traditional arithmetic, is multivalent, even when confined to denumerable models. Nevertheless, the opinion quoted above is not absurd, as evidenced, for example, by second-order logic, in which quantification is permitted not only over elements of a given base but also all relations defined on the base (normal second-order models). In this calculus Peano arithmetic becomes univalent, for the induction axioms are "merged", as it were, into a single second-order axiom. True, the second-order calculus, though interesting from the theoretical standpoint, is of no practical use: it is incomplete and so one cannot even enumerate the formulas deducible from one given formula (see Volume 1, Chapter 6).

After multivalence comes the turn of *refinement*. According to this criterion, a mathematician desirous of being "modern" should refine his axioms, reducing them to the weakest possible postulates still permitting development of the theory. Two elementary examples will suffice. Before (and with a view to) establishing the Bolzano-Weierstrass theorem: "any bounded ω-sequence of real numbers possesses a convergent ω-subsequence", we are told to prove first that "any ω-sequence with totally ordered values has a decreasing, increasing or constant subsequence". This separation of topological considerations from considerations of order yields a gain in clarity; there is also a gain in mathematical interest, for the suggested proposition is a special case of Ramsey's theorem (1926), which is fundamental in combinatorial theory and the theory of relations (see Volume 1, Chapter 3, Exercise 4). Another example: before defining the traditional concept of a metric space, in which the distances are real numbers, it is supposed to be better (at least, as a first step) to stipulate merely that the distances consitute a totally ordered additive group. Thanks to the triangle inequality, the intersection of two open balls is a union of open balls, and so this weaker version of a metric space is a topological space. One can even develop an interesting theory with this weak axiom system, including spaces with transfinite distances (integers or rational numbers, according to Hessenberg, 1906), which transcend the bounds of the traditional theory.

Another widely held view is that the modern mathematician should proceed systematically from the general to the particular. Teachers are well aware of the cost incurred by this procedure, which gives rise to complications and obscurities. If many of them resign themselves to its use, this is apparently in order to appease their logical conscience; it would be proper to undeceive them. The logical point of view pictures mathematics as a multitude of theories, some of which are representable in others. This "representability" is the sole relation that introduces some order and hierarchy among the theories. It is meaningful to say that Peano arithmetic is representable in Zermelo set theory but not in Tarski's theory of the real field. On the other hand, it is meaningless to say that the notion of a natural number or a Zermelo set logically precedes that of a real number, or vice versa; there is no priority in mathematics, neither is there any privileged logical order in which the concepts should be introduced.

If generality has any mathematical meaning, this may only be in relation to representability: one axiom system is more general than another if the former can represent the latter, either directly, as in the case of geometries (via definitions and theorems which interpret the desired primitive notions and axioms), or less directly, as in the case of axiomatic set theory (where the axioms of set theory without the axiom of choice enable one to define constructible sets and to prove that these constitute a model of set theory plus the axiom of choice; see Gödel, 1940); and the representability may be even less direct. Now it turns out that the most general theories are also the most complicated, the most recent and the most liable to revision. If one proceeds systematically from the general to the particular, an ideal treatise on the elements of mathematics must postpone arithmetic and combinatorial theory to the last chapters, relegate Cantor set theory (Zermelo-Fraenkel or von Neumann-Bernays-Gödel) to the second half, and discuss inaccessible sets in the first half. Above all, one would have to recast the first chapter monthly in order to allow for the latest demand of the ultra-categoricians, all this in the knowledge that even if this is done satisfactorily, a new, stronger and more urgent demand will be presented even before the provisory version is completed.

Since logic imposes no order among mathematical theories, the mathematician drawing up a teaching project or a treatise on the elements of mathematics should be guided first and foremost by practical and pedagogical considerations. Up to the present, only a few axiom systems have "proved themselves" under the handling of mathematicians or logicians; we are referring to Peano arithmetic, the theory of real fields (Tarski, 1940 and 1951) (the simplest representation of elementary geometry), and finally the various set-theoretic axiomatic systems, whose role as a unifying factor in mathematics has been greatly exaggerated (in actual fact the very opposite is true). The simplest set theory, which should precede the others in any pedagogically inspired approach, is combinatorial theory or the theory of finite sets, which is simply Zermelo-Fraenkel theory with the axiom of infinity replaced by an axiom of finiteness stating that every set is finite, say in the sense of Tarski (1924): for any a and any set of subsets b of a, at least one subset b is minimal under inclusion (see Volume 1, Chapter 1, Exercise 1).

To complicate the situation, the first two of the above-mentioned axiom systems possess a certain shortcoming. The Peano system does not easily lend itself to the introduction of finite fields, p-adic numbers and quadratic forms, which seem indispensable for the development of arithmetic (see, e.g., the Appendix to this volume). For the moment, therefore, one should reserve this axiom system for beginners, who will find it an excellent exercise of logic to conduct a critical survey of the usual inductive definitions, replacing them by logically correct definitions, as suggested by Gödel (1930; see Volume 1, 7.2.2). As to the axiomatic theory of real fields, though known to be decidable, it is quite weak. For example, it excludes any appeal to the general concept of a natural number: the definition of a polygon, the theorem stating that two triangles of equal area may be divided into finitely many pairwise congruent triangles (see below, 3.6). On the other hand, the axiomatic set theories by no means possess the perspicuity and simplicity sometimes attributed to them. Thus, the axioms of induction, which they claim to avoid by considering sets of integers, reappear on a higher level, for example, in the form of the following accessibility axiom: "If a property P is true for the set of natural numbers and is preserved under passage to subsets, power sets, unions and images under mappings, then P is true for any set."

Philosophers frequently query the possibility of defining the beauty of a theorem or, if so desired, esthetics or "good taste" in mathematics. Of course, when the "technician" of logic lays hold of such a philosophical or psychological concept, he transforms it into a rigorous concept, but in so doing he deprives it of much of its generality and philosophical interest. For example, he might try, as a first approximation, to measure the beauty of a theorem by the ratio between the length of its shortest possible proof and the length of the theorem itself. The mathematician deems himself esthetically gratified when he succeeds in proving a short statement by way of a few dozen lemmas, each requiring a proof of two or three pages. Moreover, the statment should not turn out to be an easy corollary or unnoticed form of a previously known statement, in which case its beauty is merely a reflection. However, the length of a theorem or proof is not a rigorous notion unless one stipulates the concepts allowed to figure therein; the simplest approach is to allow only the primitive concepts of the axiom system with which one is working, invoking none of the abbreviations furnished by definitions.

We conclude with a few words on existence in mathematics. In the formalist view, mathematical "beings" are solely a human invention, and the logical laws that they obey are the rules of a game of the intellect. In the platonist view, mathematical beings are discovered, in the same way as physical laws. They exist *a priori* in the world of ideas, as the laws of physics exist in the universe. Nevertheless, for mathematicians of either conviction, the criterion for mathematical existence is consistency, or noncontradiction. For the formalists consistency defines mathematical existence, since it is the precondition for the game to function; for the platonists, consistency is guaranteed by the presence of the being in the world of ideas.

Despite its metaphysical aspect, the platonist doctrine is seductive for the logician, even if he is a materialist or an agnostic. Indeed, it reproduces on a higher level the connections between the represented theory (mathematics as conceived by man) and the representing metatheory (the world of ideas). It may well be asked whether the formalist and platonist tendencies are really opposed. Even the formalist is inclined to believe that mathematical concepts and theorems, though we may not know how to display them in any form other than a game of the human intellect, probably exist among thinking beings quite different from man. It does not even seem necessary to stipulate living beings, beings which assimilate, reproduce, possess a will or an adapative capacity; a self-programming computer is in a sense a thinking being. The appearance of mathematical concepts and their subjection to logical laws would seem to be a general phenomenon, distributed throughout the universe. The universe is not only physical, chemical, biological, psychological, but also "mathematico-logical": these thinking beings reveal the existence of a mathematics pervading the entire universe. We are thus led back to the conception of a world of ideas linked to the physical world; before adopting either viewpoint, formalism or platonism, we should wait for more clear-cut indications that these viewpoints are indeed incompatible.

CHAPTER 1

LOCAL ISOMORPHISM AND LOGICAL FORMULA; LOGICAL RESTRICTION THEOREM

1.1. (k, p)-ISOMORPHISM

Let R and R′ be two multirelations with bases E and E′, respectively, and f a local isomorphism of R onto R′, defined on a subset $F \subseteq E$ and mapping it onto some $F' \subseteq E'$ (see Volume 1, 4.1). We associate with f certain pairs of integers (k, p), defined recursively as follows.

Every local isomorphism is an $(0, p)$-*isomorphism* for any integer $p \geqslant 0$.

If $k \geqslant 1$, the mapping f is a (k, p)-*isomorphism* if the following condition holds: for any set $\bar{F}$ obtained by adjoining q elements of E to F $(0 \leqslant q \leqslant p)$, there exists an extension $\bar{f}$ of f to $\bar{F}$ which is a $(k-1, p-q)$-isomorphism of R onto R′; the same holds true with F, E, f, R, R′ replaced by F′, E′, f^{-1}, R′, R, respectively.

A bijective mapping f of one subset of E onto another is a (k, p)-*automorphism* of R if it is a (k, p)-isomorphism of R onto itself.

Example. Let R and R′ be two chains (i.e., total orderings) with bases E and E′. A bijective mapping f of a subset F of E onto a subset F′ of E′ is a local isomorphism if and only if it preserves the order of the elements. Let F be finite; then f is a $(1, 1)$-isomorphism if and only if it preserves order and, moreover, maps any two R-consecutive elements onto R′-consecutive elements and the minimal (maximal) element of R onto the minimal (maximal) element of R′; the same holds for f^{-1} with R and R′ interchanged.

In this example, let us use the term *corresponding intervals* for an R-interval whose endpoints are two elements a, b (or the interval of elements preceding a or those following a) and the R′-interval whose endpoints are $f(a)$, $f(b)$ (or the interval preceding $f(a)$ or following $f(a)$). If we omit the endpoint(s) of these intervals, f is a $(1, 1)$-isomorphism if and only if any two corresponding intervals are either both empty or both nonempty. In general, f is a $(1, p)$-isomorphism if and only if any two corresponding intervals either have the same number $<p$ of elements, or have $\geqslant p$ elements.

1.1.1. *Any (k, p)-isomorphism is also a (k', p')-isomorphism for $k' \leqslant k$ and $p' \leqslant p$.*

Any local isomorphism is a $(k, 0)$-isomorphism for any $k \geqslant 0$.

Any (p, p)-isomorphism is also a (k, p)-isomorphism for any k.

It follows that the only values of k of interest in this context are $k \leqslant p$.

1.1.2. *If a mapping f, defined on* F, *is a (k, p)-isomorphism of* R *onto* R′, *the same holds for any restriction of f to a subset of* F.

An isomorphism f of R *onto* R′ *(defined on the entire base of* R *and mapping it onto that of* R′*) is a (k, p)-isomorphism for any k and p.*

Consequently:

Any local isomorphism of R *onto* R′ *which is extendible to an isomorphism of* R *onto* R′ *is a (k, p)-isomorphism for all k and p.*

For any subset F *of the base of* R, *the identity on* F *is a (k, p)-automorphism of* R *for all k and p.*

In relation to Volume 1, 4.1.4, note the following:

Let f be a local isomorphism of an m-ary relation R onto another m-ary relation R′. Even if every restriction of f to at most m elements of the base is a (k, p)-isomorphism, f itself need not be a (k, p)-isomorphism. For example, let R and R′ be unary relations on the natural numbers, $R(x) = +$ for $x = 0, 1, 2$, $R(x) = -$ for $x \geqslant 3$, $R'(x) = +$ for $x = 0, 1$, $R'(x) = -$ for $x \geqslant 2$. Let f be the identity on $\{0, 1\}$. The restrictions of f to the singletons $\{0\}$ and $\{1\}$ are $(1, 1)$-isomorphisms. But f itself is not a $(1, 1)$-isomorphism, since it cannot be extended to a local isomorphism defined on $\{0, 1, 2\}$.

1.1.3. *If f is a (k, p)-isomorphism of* R *onto* R′, *then f^{-1} is a (k, p)-isomorphism of* R′ *onto* R. *If moreover g is a (k, p)-isomorphism of* R′ *onto* R″, *then gf is a (k, p)-isomorphism of* R *onto* R″.

1.1.4. *Let R_i $(i = 1, \ldots, h)$ denote the relations of* R *and R'_i those of* R′. *Then any (k, p)-isomorphism of* R *onto* R′ *is also a (k, p)-isomorphism of R_i onto R'_i for each i.*

In general, if N *is some sequence of the relations R_i and* N′ *the sequence of similarly indexed relations R'_i, then any (k, p)-isomorphism of* R *onto* R′ *is a (k, p)-isomorphism of* N *onto* N′.

On the other hand, if f is a (k, p)-isomorphism of R_i onto R'_i for each

$i = 1, \ldots, h$, *it does not necessarily follow that* f *is a* (k, p)*-isomorphism of* $(R_1, \ldots, R_h)$ *onto* $(R'_1, \ldots, R'_h)$ (compare Volume 1, 4.1.1).

For example, consider (R, S) and (R′, S′), where $R = R' = S =$ the ordering of the natural numbers, and $S' =$ the ordering obtained from R by permuting the integers 1 and 2. Let f be the identity mapping on $\{0\}$. Then since f is extendible to an automorphism of R, it is a (k, p)-isomorphism of R onto $R' = R$ for all k, p. Similarly, since f is extendible to an isomorphism of $R = S$ onto S′, it is a (k, p)-isomorphism of S onto S′ for all k, p. Nevertheless, f is not a (1, 2)-isomorphism of (R, S) onto (R′, S′). In fact, we have $F = F' = \{0\}$; let $\bar{F} = \{0, 1, 2\}$, and suppose there exist two integers x, y such that the mapping g defined by $g(0) = 0$, $g(1) = x$, $g(2) = y$ is a local isomorphism of (R′, S′) onto (R, S).

Since 1 precedes 2 in R′ and follows 2 in S′, this would imply that x precedes y in R and follows y in $S = R$ – contradiction.

1.1.5. Case of unary multirelations. Given a unary multirelation $R = (R_1, \ldots, R_h)$ with base E, we define the *classes associated with* R to be the subsets of E on each of which all relations R_i $(1 \leqslant i \leqslant h)$ take on the same value; there are at most 2^h such classes. Given R and another unary multirelation $R' = (R'_1, \ldots, R'_h)$ with base E′, we say that a class associated with R and a class associated with R′ *correspond* if, for every i, the relations R_i and R'_i take on the same value on the respective classes (see Volume 1, 5.5.3).

Let f *be a bijective mapping of a subset* F *of the base of* R *onto a subset* F′ *of the base of* R′. *Suppose that every element of* F *is mapped onto an element of the corresponding class, and, moreover, the sets obtained by deleting all elements of* F *and* F′ *from any two corresponding classes are either of the same cardinality* $<p$ *(where* p *is a natural number) or both of cardinality* $\geqslant p$. *Then* f *is a* (k, p)*-isomorphism of* R *onto* R′ *for any* k.

▷ If $k = 0$ or $p = 0$, a (k, p)-isomorphism is simply a local isomorphism, and the assertion is obvious. Let $k \geqslant 1$, $p \geqslant 1$. Suppose the assertion true for $k-1$ and all natural numbers $\leqslant p$; we shall prove that it holds for k and p. Extend f to $q \leqslant p$ new elements from the base of R. We obtain corresponding classes which either have the same cardinality $<p-q$ or two cardinalities $\geqslant p-q$, so that the extension of f is a $(k-1, p-q)$-isomorphism of R onto R′. ◁

1.1.6. CASE OF DISCRETE CHAINS. A chain (total ordering) is said to be *discrete* if every element other than the maximal element has an immediate successor and every element other than the minimal element has an immediate predecessor.

We distinguish four classes of discrete chains over infinite bases: chains with a minimal but no maximal element (e.g., the chain of natural numbers); chains with a maximal but no minimal element (e.g., the chain of negative integers); chains with neither maximal nor minimal element e.g., the chain of integers); and infinite chains with both minimal and maximal element (e.g., the chain of natural numbers followed by the negative integers).

Let R and R′ be two chains of the same class, over bases E and E′, and f a local isomorphism of R onto R′, mapping a finite subset $F \subseteq E$ onto $F' \subseteq E'$. Consider the corresponding intervals, minus endpoints, introduced in 1.1., and let k, p be two integers. Then:

If any two corresponding intervals are either of the same cardinality $<(p+1)^k-1$ *or both of cardinalities* $\geqslant(p+1)^k-1$, *then* f *is a* (k, p)-*isomorphism of* R *onto* R′.

▷ Obvious for $k=0$ and any p. Let $k\geqslant 1$; we suppose the assertion true for $k-1$ and all natural numbers $\leqslant p$, and prove it true for k, p. If we subdivide any discrete interval containing at least $(p+1)^k-1$ elements by means of $q\leqslant p$ elements, we obtain at least one subinterval in which the number of elements is at least $(p+1)^k-p-1$ divided by $p+1$, hence at least $(p+1)^{k-1}-1$. We can therefore define an extension of f to q new elements, satisfying the assumptions of the theorem with k and p replaced by $k-1$ and p, hence *a fortiori* with k and p replaced by $k-1$ and $p-q$. The induction hypothesis now tells us that the extension of f is a $(k-1, p-q)$-isomorphism. ◁

1.1.7. CASE OF SUCCESSION RELATIONS. A binary relation C with base E is said to be a *succession relation* if, for any element a of E, there exists at most one element b such that $C(a, b)=+$ and at most element c such that $C(c, a)=+$; these elements are assumed to be distinct from a but not necessarily from each other. We call b the *successor* of a and c the *predecessor* of a. An element with no predecessor is said to be C-*minimal*, an element with no successor C-*maximal.* A *finite interval* is a nonempty finite set obtained by starting with an element (the *initial endpoint*), and going from each element to its successor until another element (the *final endpoint*) is reached. In the cases to be considered below, the initial endpoint will never be the successor of the final endpoint, so that restrictions of C to two intervals of the same finite cardinality are always isomorphic. It is quite obvious how one defines a *denumerable interval*, which may have only one (initial or final) endpoint or no endpoints at all. One can also define a *cycle*, which has at least two elements and reproduces the same element after finitely many passages from element to successor. The elements of E may be partitioned into *succession classes*, two elements stipulated to be equivalent if they are the endpoints of some finite interval or if they form a two-element cycle. Each class is either a finite interval, an infinite interval, a cycle, or singleton.

An important special case occurs when R is a chain and we consider the succession relation *associated* with R: an element b is the successor of a if it is distinct from a, follows it in the ordering R and there is no element between a and b in R.

Let C and C′ be two succession relations on infinite bases E and E′, each containing at least one denumerable interval and only finitely many maximal and minimal elements. Let f be a bijective mapping of a finite subset F on E onto a finite subset F′ of E′, k and p two natural numbers. The elements of F are partitioned into finitely many disjoint finite intervals, which we call C-*fragments*, and similarly the elements of F′ are partitioned into C′-fragments. Suppose that the C-fragments and C′-fragments are in bijective correspondence, with the following four conditions satisfied:

(1) Any two corresponding fragments are finite intervals of the same cardinality; the local isomorphism of C onto C′ which maps one fragment onto the other is the restriction of f to C.

(2) Every minimal (maximal) element of C is in some C-fragment; the same holds for C′ and the corresponding C′-fragments.

(3) Every cycle contains at least $(p+1)^k$ elements.

(4) Two distinct C-fragments are either contained in two distinct succession classes, or else the interval or two intervals extending from the final endpoint of one to the initial endpoint of the other contains at least $(p+1)^k$ elements (not counting these endpoints). The same holds for the

interval extending from the final to the initial endpoint of such a C-fragment contained in a cycle. The same condition holds for C′.

Under these assumptions, the mapping f is a (k, p)*-isomorphism of* C *onto* C′.

▷ First let $k=0$, and note that then the final endpoint of a C-fragment cannot be the predecessor of the initial endpoint of a C-fragment, for there must be at least one other element between the two. The same holds for C′. It follows that combination of the isomorphisms of C-fragments onto the corresponding C′-fragments yields a local isomorphism of C onto C′, in other words, a $(0, p)$-isomorphism.

Now let $k \geqslant 1$ and suppose the statement true up to $k-1$ and for all p. Let p be an integer such that conditions (1) through (4) hold for k and p. Adjoin to F at most p elements a of E. Now, if a is a term in a sequence of other elements a which can be "joined" to one of the endpoints of a C-fragment by intervals containing less than $(p+1)^{k-1}$ elements, we extend the C-fragment to contain a, extending the corresponding C′-fragment by isomorphism, thus mapping a onto the required element a'. Note that there are always at least $(p+1)^{k-1}$ elements between the most distant element a following a final endpoint and the most distant one preceding an initial element, so that condition (4) holds for the new fragments, with k replaced by $k-1$. Note also that we can always extend the isomorphism of C onto C′ without ever being obstructed by a maximal or minimal element, since the latter are already in the C′-fragments by condition (2). Finally, if a does not fall into the above category, all such elements a are collected into new C-fragments with the interval between them containing at least $(p+1)^{k-1}$ elements; condition (4) for $k-1$ holds here by virtue of conditions (3) and (4) for k. We can then associate a new C′-fragment with each of these fragments; this is possible because both C and C′ contain a denumerable interval. ◁

1.2. (k, p)-EQUIVALENCE

We consider systems of the type $(\mathrm{R}; a_1, \ldots, a_n)$, where R is a μ-ary multirelation and $a_1, \ldots, a_n$ an n-tuple of elements from the base of R (μ and n will remain fixed). A system $(\mathrm{R}'; a'_1, \ldots, a'_n)$ is said to be (k, p)*-equivalent* to $(\mathrm{R}; a_1, \ldots, a_n)$ if the mapping taking each a_i onto a'_i $(i=1, \ldots, n)$ is a

(k, p)-isomorphism of R onto R′. That this is indeed an equivalence relation (reflexive, symmetric and transitive) follows from 1.1.2 and 1.1.3.

1.2.1. Given a sequence of integers μ and integers n, k, p, we have the following proposition:

There are only finitely many (k, p)-equivalence classes of systems of the above type.

▷ If $k=0$, (k, p)-isomorphism is simply local isomorphism. The equivalence class of $(R; a_1, \ldots, a_n)$ is completely determined by the restriction of R to the set of elements a_i $(i=1, \ldots, n)$, hence also by the image of this restriction under the bijective mapping which takes a_1 to 1, a_2 to 1 or 2 according as $a_2=a_1$ or $a_2 \neq a_1$, and so on. It follows that the equivalence class is determined by a multirelation of arity μ whose base is a subset of $\{1, \ldots, n\}$; there are only finitely many such multirelations.

Now let $k \geqslant 1$; assuming the statement true for $k-1$, any number $\leqslant p$ and any n, we shall show that it is true for n, k, p. Let $C(n, k, p)$ denote the (k, p)-equivalence classes of systems of type $(R; a_1, \ldots, a_n)$. Consider all the sequences $b_1, \ldots, b_q$ of elements of the base of R, $q \leqslant p$; with each such sequence we associate the class $C(n+q, k-1, p-q)$ which contains $(R; a_1, \ldots, a_n, b_1, \ldots, b_q)$. By assumption, given q, there are finitely many such classes. Now, by the definition of (k, p)-isomorphism (1.1), two systems of type $(R; a_1, \ldots, a_n)$ are (k, p)-equivalent if and only if they are associated with the same set of classes $C(n+q, k-1, p-q)$ for each $q=0, 1, \ldots, p$. Hence there exists a bijective mapping taking each class $C(n, k, p)$ onto a set of classes $C(n+q, k-1, p-q)$, where q ranges over the finite set $0, 1, \ldots, p$. It follows that there are only finitely many classes $C(n, k, p)$. ◁

1.2.2. Particular case: (k, p)-equivalence of multirelations. We shall say that two multirelations R, R′ of the same arity are *(k, p)-equivalent* if the empty mapping is a (k, p)-isomorphism of R onto R′.

A sufficient condition for R *and* R′ *to be (k, p)-equivalent is that there exist a (k, p)-isomorphism of* R *onto* R′ (this condition is trivially necessary). Indeed, the restriction of this (k, p)-isomorphism to the empty set is a (k, p)-isomorphism, by 1.1.2.

Examples. Two discrete chains with infinite bases, both in the same class of those defined in 1.1.6, are (k, p)-equivalent for any k and p, hence logically equivalent by 1.3.3 below (see [TAR, 1934]). The same holds for two succession relations associated with two chains of the same class (see 1.1.7).

Two unary multirelations of the same arity such that any two corresponding classes are either of the same cardinality $<p$ or both of cardinalities $\geqslant p$ are (k, p)-equivalent for any k (see 1.1.5).

1.2.3. Given a sequence of integers μ and two integers k, p:

There are only finitely many (k, p)*-equivalence classes of* μ*-ary multirelations* (this is a particular case of 1.2.1).

1.2.4. R *and* R′ *are* $(1, p)$*-equivalent if and only if every restriction of* R *to at most* p *elements is isomorphic to a restriction of* R′ *and conversely.*

Examples. Two chains defined on $\geqslant p$ elements are $(1, p)$-equivalent. Two infinite chains are $(1, p)$-equivalent for any p.

Any relation which is $(1, 3)$-equivalent to a chain is itself a chain. Indeed, if the chain R in question is defined on one or two elements, any relation which is $(1, 3)$-equivalent to R is also isomorphic to R. If R is defined on three elements or more and R′ is $(1, 3)$-equivalent to R, then any restriction of R′ to one, two or three elements is a chain, so that R′ is reflexive, transitive and antisymmetric and any two elements in its base are comparable. Thus R′ is a chain.

1.2.5. *If* R *is defined on* r *elements, any multirelation which is* $(1, r+1)$*-equivalent to* R *is isomorphic to* R.

1.2.6. A binary relation R is said to be *functional* if for any x in its base there is exactly one y such that $R(x, y) = +$.

Any relation which is $(2, 3)$*-equivalent to a functional relation is a functional relation.*

▷ Let R be the given functional relation and E its base. Let R′ be a relation with base E′ which is $(2, 3)$-equivalent to R. For any $x' \in E'$, there exists a $(1, 2)$-isomorphism f of R′ onto R, defined on $\{x'\}$. Let $x = f(x')$; there exists y such that $R(x, y) = +$. There exists a local isomorphism g of R onto R′ which is an extension of f^{-1} and is defined on $\{x, y\}$. Setting $y' = g(y)$, we have $R'(x', y') = +$. We claim that y' is the only element with this property. Let x', y', z' be such that $R'(x', y') = R'(x', z') = +$; there exists a $(1, 0)$-isomorphism, hence also a local isomorphism, of R′ onto

R, defined on $\{x', y', z'\}$ and mapping these elements onto x, y, z in such a way that $R(x, y) = R(x, z) = +$. It now follows by our assumptions on R that $y = z$, so that $y' = z'$. ◁

1.2.7. Let R, R′ be two chains with bases E, E′ and f a local isomorphism of R onto R′, with domain F and range $F' = f(F)$. Since the domain and range may be infinite, we generalize the concept of corresponding intervals (see 1.1) as follows. Define an *F-interval* to be an interval of R consisting of all the elements of E − F which lie in the same position relative to each element a of F (all preceding a or all following a). In other words, we take a cut in the chain $R \mid F$ and consider all the elements of E − F in this cut. If we associate with each element a of F its image $f(a)$ in F′, each cut of $R \mid F$ is thus mapped via f onto a certain cut, so that for each F-interval we have a *corresponding* F′*-interval*, defined by the image cut.

A local isomorphism is a (k, p)-isomorphism of R *onto* R′ *if and only if the corresponding intervals are pairwise (k, p)-equivalent.*

▷ The assertion is true for $k = 0$, because all chains, including the binary relation with empty base, are $(0, p)$-equivalent for any integer p. Now suppose the assertion true for $k-1$ $(k \geqslant 1)$ and every p; we shall prove it true for a local isomorphism f of R onto R′ with integers k and p.

Suppose that f is a (k, p)-isomorphism. For any sequence of q elements $(q \leqslant p)$ chosen from an F-interval A, we have a sequence of q elements in the corresponding F′-interval A′, and the extension $\bar{f}$ of f obtained by adding these pairs of elements is a $(k-1, p-q)$-isomorphism of R onto R′. Thus any subinterval of A and the corresponding subinterval of A′ are by assumption $(k-1, p-q)$-equivalent. It follows that the restriction of f to the added elements is a $(k-1, p-q)$-isomorphism of A onto A′. Interchanging A and A′, we show similarly that the intervals are (k, p)-equivalent.

Conversely, suppose that corresponding intervals are pairwise (k, p)-equivalent. Given any sequence of q elements of the base of R, we partition them into subsequences, one in each F-interval. For every such subinterval we have a sequence with the same number of terms in the corresponding F′-interval, and the element pairs thus determined define a $(k-1, p-q)$-isomorphism of the F-interval onto the corresponding F′-interval. By hypothesis, the corresponding subintervals are pairwise

$(k-1, p-q)$-equivalent. Thus the extension of f obtained by adding the above element pairs is a $(k-1, p-q)$-isomorphism of R onto R′. The same reasoning, with R and R′ interchanged, shows that f is a (k, p)-isomorphism of R onto R′. ◁

1.3. Characteristic of a logical formula. Relations between (k, p)-isomorphism and logical formula

With any logical formula P we now associate a pair of integers (k, p), $k \leqslant p$, called the *characteristic* of P and defined as follows:

If P is free, $k=p=0$.

If $P=\alpha\, P_1 \dots P_h$, where α is an h-ary connection and P_i is a logical formula with characteristic (k_i, p_i) $(i=1,\dots, h)$, then the characteristic of P is (k, p), where $k=\max(k_1,\dots, k_n)$, $p=\max(p_1,\dots, p_h)$.

If $P= \underset{i_1,\dots,i_r}{\forall} Q$ or $\underset{i_1,\dots,i_r}{\exists} Q$ and the characteristic of Q is (k, p), then the characteristic of P is $(k+1, p+r)$, where r is the number of indices $i_1,\dots, i_r$ of the quantifier.

Examples. The formula $\underset{1}{\forall} x^1 x^1$, which represents the class of reflexive relations, has characteristic $(1, 1)$. The formula $\underset{3}{\forall} \rho x^3 x^1 \bigvee \rho x^2 x^3$ has characteristic $(1, 1)$. The formula

$$\underset{1,2,3}{\forall} (\rho x^1 x^2 \bigwedge \rho x^2 x^3) \Rightarrow \rho x^1 x^3,$$

which represents the class of transitive formulas, has characteristic $(1, 3)$. The formula

$$\underset{1}{\forall} \underset{2}{\exists} \rho x^1 x^2 \bigwedge \underset{1,2,3}{\forall} (\rho x^1 x^2 \bigwedge \rho x^1 x^3) \Rightarrow x^2 \equiv x^3,$$

which represents the class of functional relations, has characteristic $(2, 3)$.

For any two characteristics, we write $(k', p') \leqslant (k, p)$ when $k' \leqslant k$ and $p' \leqslant p$.

1.3.1. *Let* R, R′ *be two μ-ary multirelations, f a (k, p)-isomorphism of* R *onto* R′, *and* P *a μ-ary formula of characteristic $\leqslant (k, p)$. Then f is a local isomorphism of the relation* P(R) *onto* P(R′).

In particular, given a quantifier $\mathcal{Q} = \underset{i_1,\dots,i_r}{\forall}$ or $\mathcal{Q} = \underset{i_1,\dots,i_r}{\exists}$, where $i_1,\dots, i_r$

are r indices, any $(1, r)$-isomorphism of R onto R′ is a local isomorphism of $\mathscr{Q}(R)$ onto $\mathscr{Q}(R')$.

▷ The assertion is true when P is free, for any local isomorphism of R onto R′ is then a local isomorphism of P(R) onto P(R′).

If $P=\alpha P_1 \dots P_h$, each formula P_i $(i=1,\dots, h)$ is of characteristic $\leqslant (k, p)$. Suppose our assertion true for each P_i; then f is a local isomorphism of $P_i(R)$ onto $P_i(R')$ for each $i=1,\dots, h$, hence also of P(R) onto P(R′) (since the connector defined by α is a free operator). If $P = \underset{i_1,\dots,i_r}{\forall} Q$, where $i_1,\dots, i_r$ are r integers, the characteristic of the formula Q is $\leqslant (k-1, p-r)$. We may assume without loss of generality that $i_1 = n+1,\dots, i_r = n+r$, where n is the arity of P and $n+r$ that of Q; suppose the assertion true for Q. Let $a_1,\dots, a_n$ be n elements (not necessarily distinct) in the domain of definition of f such that $P(R)(a_1,\dots, a_n) = +$. On the one hand, for any $b_1,\dots, b_r$ in the base of R, we have

$$Q(R)(a_1,\dots, a_n, b_1,\dots, b_r) = +.$$

On the other hand, for any $b'_1,\dots, b'_r$ in the base of R′, there exist elements $b_1,\dots, b_r$ such that the extension $\bar{f}$ of f taking each a_i to a'_i $(i=1,\dots, n)$ and each b_j to b'_j $(j=1,\dots, r)$ is a $(k-1, p-r)$-isomorphism. By the induction hypothesis, this extension $\bar{f}$ is a local isomorphism of Q(R) onto Q(R′), so that

$$Q(R')(a'_1,\dots, a'_n, b'_1,\dots, b'_r) = +,$$

and since this is true for all $b'_1,\dots, b'_r$, we obtain

$$P(R')(a'_1,\dots, a'_n) = +.$$

Interchanging R and R′, one now proves in similar fashion that if $P(R)(a_1,\dots, a_n) = -$, then

$$P(R')(a'_1,\dots, a'_n) = -.$$

Varying the elements $a_1,\dots, a_n$, we thus conclude that f is a local isomorphism of P(R) onto P(R′). ◁

1.3.2. *Let* R *be a μ-ary multirelation and $a_1,\dots, a_n$ a sequence of n elements of its base; let k, p be given integers. Then there exists a (μ, n)-ary formula* P *of characteristic $\leqslant (k, p)$ such that* $P(R')(a'_1,\dots, a'_n) = +$ *if and only if* $(R'; a'_1,\dots, a'_n)$ *is (k, p)-equivalent to* $(R; a_1,\dots, a_n)$.

▷ If $k=0$, we let P be the operator $\mathscr{P}$ such that $\mathscr{P}(R')(a'_1,\ldots,a'_n)=+$ when $(R'; a'_1,\ldots,a'_n)$ is (0, 0)-equivalent to $(R; a_1,\ldots,a_n)$, i.e., when the mapping taking each a_i to a'_i $(i=1,\ldots,n)$ is a local isomorphism of R onto R′. The operator $\mathscr{P}$ is free (see Volume 1, 4.3.1).

If $k\geqslant 1$, suppose the assertion true for $k-1$ and all integers $\leqslant p$. We must prove it for k and p. Given $q\leqslant p$, consider the $(k-1, p-q)$-equivalence classes C of systems $(R; a_1,\ldots,a_n, b_1,\ldots,b_q)$, where $b_1,\ldots,b_q$ are q arbitrary elements of the base of R. Let C′ denote the remaining $(k-1, p-q)$-equivalence classes (i.e., other than the classes C) of systems comprising a μ-ary relation and a sequence of $n+q$ elements. Now, the number of classes C is finite (1.2.1), as is the number of classes C′; by assumption, for each C we have a formula Q and for each C′ a formula Q′, these formulas being $(\mu, n+q)$-ary and of characteristic $\leqslant(k-1, p-q)$, and such that

$$Q(R')(a'_1,\ldots,a'_n, b'_1,\ldots,b'_q)=+$$

if and only if the system $(R'; a'_1,\ldots,a'_n, b'_1,\ldots,b'_q)$ is in C (or in C′). Now form all the formulas $\underset{n+1,\ldots,n+q}{\exists} Q$ and $\neg \underset{n+1,\ldots,n+q}{\exists} Q'$ for all Q and Q′. These are (μ, n)-ary formulas of characteristic $\leqslant(k, p)$, and their conjunction P is the required formula. ◁

1.3.3. *Let* R, R′ *be two* μ*-ary multirelations,* $a_1,\ldots,a_n$ *a sequence of elements of the base of* R *and* $a'_1,\ldots,a'_n$ *a sequence of elements of the base of* R′. *Then the mapping which takes each* a_i *to* a'_i $(i=1,\ldots,n)$ *is a* (k, p)*-isomorphism of* R *onto* R′ *if and only if*

$$P(R)(a_1,\ldots,a_n)=P(R')(a'_1,\ldots,a'_n)$$

for every logical formula P *of arities* (μ, n) *and characteristic* $\leqslant(k, p)$.

This is an immediate corollary of 1.3.1 and 1.3.2. In particular:

Two μ*-ary multirelations* R, R′ *are* (k, p)*-equivalent if and only if* $P(R)=P(R')$ *for every bound formula* P *of predicarity* μ *and characteristic* $\leqslant(k, p)$.

They are (k, p)*-equivalent for all* k *and* p *if and only if they are logically equivalent* (see Volume 1, 5.6).

A mapping which is a (k, p)-isomorphism [(k, p)-automorphism] for all k and p will be called a *logical isomorphism* [*logical automorphism*].

1.3.4. CHARACTERIZATION OF LOGICAL OPERATORS AND LOGICAL CLASSES. *An operator $\mathscr{P}$ is represented by a logical formula of characteritic $\leqslant(k,p)$ if and only if it satisfies the following condition: any (k, p)-isomorphism of a multirelation* R *onto* R′ *(both assignable to $\mathscr{P}$) is a local isomorphism of $\mathscr{P}$*(R) *onto $\mathscr{P}$*(R′).

This condition may be weakened as follows: any (k, p)-isomorphism of R onto R′ *defined on at most n elements* (where n is the arity of $\mathscr{P}$) is a local isomorphism of $\mathscr{P}$(R) onto $\mathscr{P}$(R′) (Volume 1, 4.1.4).

▷ If $\mathscr{P}$ is represented by a formula of characteristic $\leqslant(k, p)$, the validity of the original condition follows from 1.3.1.

Conversely, let $\mathscr{P}$ satisfy the weakened condition. Then, given two (k, p)-equivalent systems $(R; a_1,..., a_n)$ and $(R'; a'_1,..., a'_n)$, we have

$$\mathscr{P}(R)(a_1,..., a_n)=\mathscr{P}(R')(a'_1,..., a'_1).$$

With each (k, p)-equivalence class J we can associate (see 1.3.2) a formula Q of characteristic $\leqslant(k, p)$ such that $Q(R)(a_1,..., a_n)$ takes the value $+$ or $-$ according as $(R; a_1,..., a_n)$ is in J or not in J.

Consider the classes J which assign $\mathscr{P}$ the value $+$. There are only finitely many such classes (see 1.2.1), so that we can speak of the disjunction P of the corresponding formulas Q (P may reduce to the operator $-$, as the case may be).

Then P is of characteristic $\leqslant(k, p)$ and always assumes the same value as $\mathscr{P}$, so that it represents $\mathscr{P}$. ◁

In particular, setting $n=0$, we obtain:

A class of multirelations is a logical class, and in fact represented by a bound formula of characteristic $\leqslant(k, p)$, if and only if it is a union of (k, p)-equivalence classes.

1.3.5. Let k, k', p, p' be four natural numbers and $\mathscr{P}$ a logical operator represented by a formula of characteristic $\leqslant(k, p)$. *Then any $(k+k', p+p')$-isomorphism of a multirelation* R *onto* R′ *is a (k', p')-isomorphism of $\mathscr{P}$*(R) *onto $\mathscr{P}$*(R′).

▷ This reduces to the previous assertion for $k'=p'=0$.

Let $k'\geqslant 1$ and assume the assertion true for $k'-1$ and all natural numbers $\leqslant p'$. Let f be a $(k+k', p+p')$-isomorphism of R onto R′, defined on F. Extending F to $\bar{F}$ by adjoining $q\leqslant p'$ elements, we have an extension of f to $\bar{F}$ which is a $(k+k'-1, p+p'-q)$-isomorphism of R onto R'. By hypothesis, it is also a $(k'-1, p'-q)$-isomorphism of $\mathscr{P}$(R) onto $\mathscr{P}$(R'). Interchanging R and R', we similarly prove the converse. It follows that f is the required (k', p')-isomorphism. ◁

1.4. Logical extension and logical restriction; logical restriction theorem

Let R be a multirelation with base E. An extension R' of R to $E' \supseteq E$ is said to be a *logical extension* of R if, for any finite subset F of E and any integers k, p, the identity mapping on F is a (k, p)-isomorphism of R onto R'. R is also said to be a *logical restriction* of R'.

It follows from 1.3.3 that an equivalent definition is as follows: for any finite sequence $a_1, \ldots, a_n$ of elements of E and any logical formula P of arity n to which both R and R' are assignable,

$$P(R)(a_1, \ldots, a_n) = P(R')(a_1, \ldots, a_n).$$

If the base E of R is finite, the only logical extension of R is R itself. Indeed, if R' is a logical extension of R, defined on $E' \supseteq E$, then the identity on E is a (k, p)-isomorphism for all k, p, hence a $(1, 1)$-isomorphism; however, for $x \in E' - E$ there is no bijective mapping of $E \cup \{x\}$ into E.

On the other hand, if E is infinite there exist proper logical extensions of R. Thus, let R be the chain of natural numbers and R' a discrete chain without maximal element, an extension of R, in which each element is either a natural number or follows all the natural numbers. Now, the identity mapping f on any finite set of natural numbers defines corresponding intervals which are either identical or both infinite. Thus f is a (k, p)-isomorphism of R onto R' for any k and p; in other words, R' is a logical extension of R (see 1.1.6).

As another example, the field of real numbers (the birelation consisting of the sum and product of real numbers) is a logical extension of the field of algebraic numbers [TAR, 1951] (see below, 3.6).

Any logical extension of a multirelation R *is logically equivalent to* R. Indeed, if R' is a logical extension of R, then R' is (k, p)-equivalent to R for all k and p (see 1.2.2) and therefore logically equivalent to R (see 1.3.3).

Note that R' *may be an extension of* R *and logically equivalent to* R, *but nevertheless not a logical extension of* R. For example, consider the chain R of natural numbers and its extension R′ to the natural numbers plus the integer -1, which is the minimal element of R′. Then R′ is the image of R under the isomorphism $f(x)=x-1$, and thus it is an extension of R which is logically equivalent to R. However, the identity mapping on $\{0\}$ is not a (1, 1)-isomorphism of R′ onto R, for it cannot be extended to a local isomorphism defined on $\{-1, 0\}$; thus R′ is not a logical extension of R (see Exercise 6).

1.4.1. (1) *If* R′ *is a logical extension of* R *and* R″ *a logical extension of* R′, *then* R″ *is a logical extension of* R.

(2) *Let* R′ *be an extension of* R; *if* R *and* R′ *have a common logical extension, then* R′ *is a logical extension of* R.

These assertions follow from the transitivity of (k, p)-isomorphism (see 1.1.3).

(3) *Let* R *be a multirelation with base* E *and* E_i *a family of subsets of* E *such that:*

(i) E *is the union of the subsets* E_i.

(ii) *The family* E_i *is directed: for any two subsets* E_i *and* E_j, *there is a third subset* E_k *such that* $E_k \supseteq E_i \cup E_j$.

(iii) *If* $E_i \subseteq E_j$, *then* $R \mid E_j$ *is a logical extension of* $R \mid E_i$.

Under these assumptions, R *is a logical extension of* $E \mid E_i$ *for each* i (see [TAR-VAU, 1957]).

▷ For any E_i and any finite subset F of E_i, the identity mapping on F is a local isomorphism, hence a $(0, p)$-isomorphism of $R_i = R \mid E_i$ onto R, and this is true for every integer p. Let $k \geqslant 1$; assuming that for all E_i, F, p, the identity mapping on F is a $(k-1, p)$-isomorphism of R_i onto R, we shall perform the induction step from $k-1$ to k. Suppose given a subset E_i, a finite subset F of E_i and a number p. Extend F by adjoining $q \leqslant p$ elements of E_i; then, by hypothesis, the identity mapping on this extension of F is a $(k-1, p-q)$-isomorphism of R_i onto R. On the other hand, let $\bar{F}$ be an extension of F by $q \leqslant p$ elements of E. By conditions (1) and (2) of our assertion, there exists a subset E_j containing $E_i \cup \bar{F}$. By hypothesis, the identity mapping on $\bar{F}$ is a $(k-1, p-q)$-isomorphism of R onto $R_j = R \mid E_j$. Moreover, since R_j is a logical extension of R_i, there exists a bijective mapping f of $\bar{F}$ onto a subset of E_i, coinciding with the identity on F, which is a $(k-1, p-q)$-isomorphism of R_j onto R_i. Finally, this mapping f is a $(k-1, p-q)$-isomorphism of R onto R_i, whence it follows that the identity mapping on F is a (k, p)-isomorphism of R_i onto R, proving the assertion. ◁

1.4.2. Let I, S be the chain and sum of natural numbers, I′ the chain obtained by adding "after" I the chain of all integers, which is a logical extension of I (see 1.4 and 1.1.6).

There exists no relation S′ *such that* (I′, S′) *is* (2, 6)-*equivalent to* (I, S). (Compare this with the result of Volume 1, Chapter 3, Exercise 7, concerning 1-extensions.)

▷ Suppose on the contrary that such a relation S′ exists; with any two elements a', b' of its base it associates exactly one element c' such that $S'(a', b', c') = +$. Indeed, given a' and b', there exists a (1, 1)-isomorphism of S′ onto S which maps them onto natural numbers a, b. Setting $c = a + b$ (mod S), we have $S(a, b, c) = +$, and so there exists c' such that $S'(a', b', c') = S(a, b, c) = +$. The uniqueness of c' follows from the fact that S and S′ are (1, 4)-equivalent, so that they have the same restrictions (up to isomorphism) to sets of four elements.

We claim that $S'(a', 0, a') = +$ for any element a' of the base of S′. Indeed, for any a' there exists a (1, 1)-isomorphism of (I′, S′) onto (I, S) which maps $a', 0$ onto integers a, u. Since corresponding intervals for I and I′ are either both empty or both nonempty, u is the minimal element of I, so that $u = 0$ and $S'(a', 0, a') = S(a, 0, a) = +$.

Now consider three elements a', b', c' such that $S'(a', b', c') = +$ and let a'_1 be the successor of a' (mod I′). Then the element associated by S′ with a'_1 and b' is the successor c'_1 of c' (mod I′). Indeed, by hypothesis there exists a (1, 1)-isomorphism of (I′, S′) onto (I, S) which maps a', b', c', a'_1, c'_1 onto certain integers a, b, c, a_1, c_1 such that $S(a, b, c) = S'(a', b', c') = +$. Since corresponding intervals are either both empty or both nonempty, a_1 is the successor of a and c_1 the successor of c (mod I), so that $S(a_1, b, c_1) = +$ and consequently $S'(a'_1, b', c'_1) = +$.

It follows from the above results that the restriction of S′ to the natural numbers coincides with S. Moreover, for each natural number u and each integer a' we have $S'(a', u, b') = +$ if and only if b' is the u-th successor of a' (mod I′).

Let a' be an integer. There exists b' such that $S'(a', a', b') = +$, and there is a local isomorphism of (I′, S′) onto (I, S) which maps a', b' onto two natural numbers a, b such that $S(a, a, b) = +$; hence $I(a, b) = +$ and so $I'(a', b') = +$. Thus b' follows a' (mod I'). There exists a natural number u such that b' is the u-th successor of a', so that $S'(a', u, b') = +$ and $u \neq a'$. Finally, there exists a local isomorphism of (I', S') onto (I, S) which maps a', b', u onto three natural numbers a^*, b^*, u^* such that $S(a^*, a^*, b^*) = +$, $S(a^*, u^*, b^*) = +$ and $u^* \neq a^*$ – contradiction. ◁

1.4.3. *Let* R *be a multirelation with base* E *and* D *a subset of* E. *Suppose that for any finite subset* F *of* E *and all integers k, p there exists a bijective mapping of* F *into* D *which maps each element of* $F \cap D$ *onto itself and is a* (k, p)*-automorphism of* R. *Then* $R \mid D$ *is a logical restriction of* R.

▷ Let G be a finite subset of D. The identity mapping on G is clearly a local isomorphism, hence a $(0, p)$-isomorphism of R onto $R \mid D$ for any p.

Let $k \geqslant 1$. Assuming that the identity mapping on any finite subset of D is a $(k-1, p)$-isomorphism of R onto $R \mid D$ for any p, we shall show that the identity mapping g on G is a (k, p)-isomorphism of R onto $R \mid D$. On the one hand, for any extension $\bar{G}$ obtained from G by adjoining $q \leqslant p$ elements of D, the identity mapping on $\bar{G}$ is an extension of g, hence, by hypothesis, a $(k-1, p-q)$-isomorphism of $R \mid D$ onto R. On the other hand, for any set F obtained from G by adding $q \leqslant p$ elements of E, there exists by hypothesis a bijective mapping f of F into D which maps each element of $G \subseteq F \cap D$ onto itself and is thus an extension of g, and which is a $(k-1, p-q)$-automorphism of R. Taking the product of this mapping with the identity mapping on $f(F)$, which is by hypothesis a $(k-1, p-q)$-isomorphism of R onto $R \mid D$, we obtain f itself, which is thus a $(k-1, p-q)$-isomorphism of R onto $R \mid D$. ◁

1.4.4. Logical restriction theorem (Refinement of the denumerable-model theorem). *Let* R *be a multirelation with infinite base* E *and* D *an infinite subset of* E. *Then there exists a set* $\bar{D}$, *equipollent to* D, *such that* $D \subseteq \bar{D} \subseteq E$ *and* $R \mid \bar{D}$ *is a logical restriction of* R (hence logically equivalent to R, by 1.4; the proof ultilizes the axiom of choice).

▷ Consider a finite sequence $s = a_1, \ldots, a_n$ of elements of D and three integers r, k, p. Let us say that two sequences of r elements of E, say $x_1, \ldots, x_r$ and $x'_1, \ldots, x'_r$, are (s, k, p)-equivalent if the systems $(R; a_1, \ldots, a_n, x_1, \ldots, x_r)$ and $(R; a_1, \ldots, a_n, x'_1, \ldots, x'_r)$ are (k, p)-equivalent, in other words, if the mapping fixing each a_i $(i=1,\ldots,n)$ and taking each x_j onto x'_j $(j=1,\ldots,r)$ is a (k,p)-automorphism of R. For given s, r, k, p, the number of equivalence classes thus generated is finite, by 1.2.1. Hence, if we vary s in D and let r, k, p range over all the natural numbers, the (infinite) set of all possible equivalence classes is equipollent to D. By the axiom of choice, there exists a function which maps each class onto a sequence $x_1,\ldots,x_r$ contained in it. Now the set D′ of terms x_j $(j=1,\ldots,r)$ of these sequences is a superset of D equipollent to D. Indeed, let s be the sequence consisting of a single element $a_1 \in \mathrm{D}$, $r=1$, $k=p=0$ and $x_1=a_1$; then the only equivalent sequence is (a_1) itself.

Repeating the preceding arguments, we define

$$\mathrm{D} \subseteq \mathrm{D}' \subseteq \mathrm{D}'' \subseteq \cdots \subseteq \mathrm{D}^{(u)} \subseteq \cdots.$$

The union $\bar{\mathrm{D}}$ of these sets $\mathrm{D}^{(u)}$ is equipollent to D and $\mathrm{D} \subseteq \bar{\mathrm{D}} \subseteq \mathrm{E}$. To complete the proof, it will suffice to show that $\bar{\mathrm{D}}$ satisfies the assumptions of 1.4.3. Indeed, let F be a finite subset of E and k, p two integers. Let $s=a_1,\ldots,a_n$ denote the sequence of elements of $\mathrm{F} \cap \bar{\mathrm{D}}$ and $x_1,\ldots,x_r$ the elements of $\mathrm{F} \cap (\mathrm{E}-\bar{\mathrm{D}})$. There exists an index u such that $\mathrm{F} \cap \bar{\mathrm{D}} \subseteq \mathrm{D}^{(u)}$ and a sequence $y_1,\ldots,y_r$ of elements of $\mathrm{D}^{(u+1)}$, hence also of $\bar{\mathrm{D}}$, which is (s,k,p)-equivalent to $x_1,\ldots,x_r$. The mapping fixing all the a_i $(i=1,\ldots,n)$ and taking each x_j onto y_j $(j=1,\ldots,r)$ is a (k,p)-automorphism of R fixing each element of $\mathrm{F} \cap \bar{\mathrm{D}}$. ◁

1.5. Examples of finitely-axiomatizable and non-finitely-axiomatizable multirelations

1.5.1. The chain of natural numbers is finitely-axiomatizable.

▷ Denote the chain in question by I. By 1.4, any discrete chain containing a minimal element and no maximal element is isomorphic to a logical extension of I and is therefore logically equivalent to I. Now these chains constitute a logical class. Indeed, since the class of all chains is universal, it is also logical (Volume 1, 4.8). To ensure the existence of the minimal element, we take the formula $\underset{x}{\exists}\underset{y}{\forall}\, \iota xy$, where the relation I may be substituted for the predicate ι. To ensure that each element have a successor (so that there is no maximal element), we take

$$\underset{x}{\forall}\underset{y}{\exists}\, x \neq y \wedge \iota xy \wedge \underset{z}{\forall}(\iota zx \vee \iota yz).$$

The existence of a predecessor for each non-minimal element is ensured by the formula

$$\underset{x}{\forall}(\underset{y}{\forall}\,\iota xy \bigvee \underset{y}{\exists}\, x \not\equiv y \bigwedge \iota yx \bigwedge \underset{z}{\forall}(\iota zy \bigvee \iota xz)).\ \lhd$$

One shows similarly that every discrete chain is finitely-axiomatizable: each of the four classes of infinite discrete chains considered in 1.1.6, with or without a minimal and/or maximal element is a logical class, in fact a logical equivalence class.

1.5.2. *Let* I *be the chain of natural numbers and* C *the succession relation* (true for pairs (x, y) such that $y = x + 1$). *Then the birelation* (I, C) *is finitely-axiomatizable.* Indeed, C is interpetable in I (see Volume 1, 7.1), and we may thus invoke Volume 1, 7.4.4.

1.5.3. THE SUCCESSION RELATION ON THE NATURAL NUMBERS IS NOT FINITELY-AXIOMATIZABLE (SEE VOLUME 1, 7.4.3).

$\rhd$ Let C be the succession relation on the natural numbers. It satisfies the following formulas, where C may be substituted for the predicate γ:

$$\underset{x}{\forall}\,\underset{y}{\exists}\,\gamma xy; \qquad \underset{xy}{\forall}\,\gamma xy \Rightarrow x \not\equiv y;$$

$$\underset{xyy'}{\forall}\,(\gamma xy \bigwedge \gamma xy') \Rightarrow y \equiv y'; \qquad \underset{xyy'}{\forall}\,(\gamma yx \bigwedge \gamma y'x) \Rightarrow y \equiv y';$$

$$\underset{u}{\exists}(\underset{x}{\forall}\,\neg\gamma xu) \bigwedge \underset{x}{\forall}\, x \not\equiv u \Rightarrow \underset{y}{\exists}\,\gamma yx,$$

and the following infinite set of formulas $(r = 1, 2, 3, \ldots)$:

$$\underset{1,2,\ldots,r}{\forall}\,(\gamma x^1x^2 \bigwedge \gamma x^2x^3 \bigwedge \cdots \bigwedge \gamma x^{r-1}x^r) \Rightarrow x^1 \not\equiv x^r.$$

These formulas are also satisfied by relations in which there exists a succession class (see 1.1.7) isomorphic to the succession relation on the natural numbers, the other succession classes being isomorphic to the succession relation on the integers. Any two such relations are logically equivalent, for by 1.1.7 the mapping taking the minimal element (0 in the case of the natural numbers) onto the minimal element is a (k, p)-isomorphism for any k and p.

To show that the succession relation C is not finitely-axiomatizable, it will suffice to show that, for any k and p, there exists a succession relation C′ which is (k, p)-equivalent to C but not logically equivalent to it. Indeed, any finite set of axioms is equivalent to a single logical formula with a certain characteristic (k, p), and any relation which is (k, p)-equivalent to C satisfies this formula (see 1.3.3). Hence this relation is logically equivalent to C. Let C′ be an extension of C defined on the set of natural numbers plus $r=(p+1)^k$ elements forming a cycle $a_1, \ldots, a_r$, i.e.,

$$C'(a_1, a_2)=C'(a_2, a_3)=\cdots=C'(a_r, a_1)=+ ;$$

where C′ takes the value – in all other cases. On the one hand, since C′ contains a cycle it does not satisfy the above axioms and is therefore not logically equivalent to C. On the other hand, by 1.1.7, the mapping of 0 onto itself is a (k, p)-isomorphism of C onto C′, so that C′ is (k, p)-equivalent to C. ◁

1.5.4. The selector which transforms each (2, 2)-ary birelation into its second relation transforms the logical equivalence class of (I, C) into the class of relations logically equivalent to C. The latter class is not a logical class. This example was anticipated in Volume 1, 5.4.3, and should be compared with the example actually given there.

The birelation (I, C) furnishes an example of multirelations M, M′ such that the concatenation MM′ is finitely axiomatizable but neither M nor M′ has this property (see Exercise 5).

1.6. (k, p)-INTERPRETABILITY

A multirelation S with the same base as a multirelation R is said to be *(k, p)-interpretable* by R if every (k, p)-automorphism of R is a local automorphism of S.

A multirelation S *is* (k, p)*-interpretable by* R *if and only if, for each relation* S_i *of* S, *there exists a logical formula* P_i *of characteristic* $\leqslant(k, p)$ *such that* $P_i(R)=S_i$. Consequently, S is interpretable by R in the sense of Volume 1, 7.1, if and only if there are two integers k, p for which S is (k, p)-interpretable by R in the above sense.

▷ The proof is similar to that in Volume 1, 4.3.5. Let $S=P(R)$, where P is a formula of characteristic $\leqslant(k, p)$. Then, by 1.3.1 with $R'=R$, every (k, p)-automorphism of R is a local automorphism of S.

Conversely, suppose that every (k, p)-automorphism of R is a local automorphism of S. Then, with any multirelation R′ of the same arity as R we can associate a relation $\mathscr{P}(R')$ as follows. Given $x'_1, \ldots, x'_n$, if there exist elements $x_1, \ldots, x_n$ in the base of R such that the mapping taking x_i onto x'_i $(i=1, \ldots, n)$ is a (k, p)-isomorphism of R onto R′, we set

$$\mathscr{P}(R')(x'_1, \ldots, x'_n) = S(x_1, \ldots, x_n).$$

This value is independent of the specific (k, p)-isomorphism chosen. If there exists no (k, p)-isomorphism of R′ onto R defined on the elements x'_i, we set

$$\mathscr{P}(R')(x'_1, \ldots, x'_n) = +.$$

It follows that every (k, p)-isomorphism of a multirelation R′ onto a multirelation R″ (both of the same arity as R) is a local isomorphism of $\mathscr{P}(R')$ onto $\mathscr{P}(R'')$. By 1.3.4, the operator $\mathscr{P}$ may be represented by a logical formula of characteristic $\leqslant (k, p)$. ◁

1.6.1. Noninterpretability proofs. As an example, we start with the chain I of natural numbers (equal to + when $x \leqslant y$), and the sum relation S (equal to + when $x+y=z$), and show that S *is not interpretable by* I (as stated in Volume 1, 7.1.3).

Given two integers k, p, let $x=(p+1)^k$ and $y=2(p+1)^k$, on the one hand, $x'=x$ and $y'=y+1$, on the other. Then the mapping taking x onto $x'=x$ and y onto y' is a (k, p)-automorphism of I, since the intervals with endpoints x, y and x', y', respectively, are both of cardinality at least $(p+1)^k-1$ (see 1.1.6). But this mapping is not a local automorphism of S, for $S(x, x, y) = +$ whereas $S(x', x', y') = -$.

It follows that S *is not interpretable in* C, where C is the succession relation ($C(x, y) = +$ when y is the successor of x). Indeed, we saw in Volume 1, 7.1, that C is interpretable by I, and so, were S interpretable in C, it would also be interpretable in I.

1.6.2. *The chain of natural numbers is not interpretable by the succession relation* (stated in Volume 1, 7.5.1).

▷ Suppose on the contrary that I is interpretable by C; then each of I and C is interpretable in the other (see Volume 1, 7.1). Since I is finitely-axiomatizable (1.5.1), the same is true of C, by Volume 1, 7.4.5. This contradicts 1.5.3. ◁

A more direct proof is as follows.

▷ Let I and C be as before, and k, p two natural numbers. Set $a=a'=0$, $b=c'=(p+1)^k+1$ and $c=b'=2(p+1)^k+2$, and let f be the mapping taking a, b, c onto a', b', c', respectively. The intervals or fragments consisting of a alone, b alone and c alone are separated by $(p+1)^k$ elements. Moreover, 0 is an element of one of them. The same holds for a', b', c'. Thus conditions 1–4 of 1.1.7 are satisfied, and so f is a (k, p)-automorphism of C. But b precedes c and b' follows c' in the chain I, so that f is not a local automorphism of I. ◁

1.6.3. We define *transitive closure* to be the operator which maps each binary relation R onto the binary relation S on the same base such that $S(x, y)=+$ whenever there exists a finite sequence of elements $x_0=x$, $x_1, \dots, x_r=y$ with

$$R(x_0, x_1)=R(x_1, x_2)=\cdots=R(x_{r-1}, x_r)=+.$$

It is evident that this operator maps the succession relation on the natural numbers onto the strict chain on the natural numbers (equal to + for (x, y) when x is strictly smaller than y).

Transitive closure is not a logical operator (stated in Volume 1, 5.2 and 5.5.5). Indeed, otherwise the strict chain of natural integers would be interpretable by the succession relation, and the same would hold for the ordinary chain (as it is freely interpretable by the strict chain), contradicting 1.6.2 above.

We can now construct a new example of *a logical operator $\mathscr{P}$ such that $\not\mathscr{S}\mathscr{P}$ is not a logical operator* (see Volume 1, 4.5, for the definition of $\not\mathscr{S}$ and compare Volume 1, 4.5.4 and 5.5.5; this example is due to G. Coray).

Consider the logical operator $\mathscr{P}$ which transforms any birelation (R, S), where R and S are binary relations, into the binary relation T defined by: $T(x_1, x_2)=+$ if either $S(x_1, x_2)=+$, or S is not transitive, or S is not deducible from R. The following formula, in which R and S are to be substituted for ρ and σ, represents the operator $\mathscr{P}$:

$$\sigma x^1 x^2 \bigvee \neg (\underset{xy}{\forall}\, \rho xy \Rightarrow \sigma xy) \bigvee \neg (\underset{xyz}{\forall}\, (\sigma xy \bigwedge \sigma yz) \Rightarrow \sigma xz).$$

Now the operator $\mathscr{P}_\sigma\mathscr{P}$ transforms (R, S) into the transitive closure of R. Let $\mathscr{Q}$ be the operator transforming R into the birelation (R, R); $\mathscr{Q}$ is logical (even free). Thus, were $\mathscr{P}_\sigma\mathscr{P}$ a logical operator, the product of $\mathscr{P}_\sigma\mathscr{P}$ and $\mathscr{Q}$, which transforms each relation R into its transitive closure, would also be a logical operator – contradiction.

1.7. Homogeneous and Logically Homogeneous Multirelations

1.7.1. Recall that a relation R is *homogeneous* if every local automorphism of R defined on finitely many elements may be extended to an automorphism of R (see Volume 1, Chapter 4, Exercise 5).

If R is denumerable and homogeneous and R′ is denumerable and (2, p)-equivalent to R for every p, then R′ is homogeneous. R′ has the same finite restrictions as R (up to isomorphism). It can be shown that R′ is isomorphic to R. In particular, if R is denumerable and homogeneous and R′ is denumerable and logically equivalent to R, then R and R′ are isomorphic.

1.7.2. A relation R with base E is said to be *logically homogeneous* if, for any finite subset F of E, there exist two integers k and p such that every (k, p)-automorphism of R defined on F may be extended to an automorphism of R. For example, a homogeneous relation (in the sense of 1.7.1) is logically homogeneous: $k=p=0$ for each F. The chain I of natural numbers is logically homogeneous: letting a be the maximal integer in F, we see that the only $(1, a+1)$-automorphism of I defined on F is the identity mapping on F, which is of course extendible to the identity mapping of the entire base. Similarly, the chain of negative integers and the chain of all integers are logically homogeneous; however, the only denumerable homogeneous chain is the chain of rational numbers (see Volume 1, Chapter 4, Exercise 5 (3)).

If two denumerable logically homogeneous relations are logically equivalent, they are isomorphic.

If R *is denumerable and logically homogeneous and* R′ *is logically equivalent to* R, *then* $R \leqslant R'$ (communicated by J.P. Calais).

Problem. In the latter case, is R isomorphic to a logical restriction of R′?

If I is the chain of natural numbers and R any relation on the natural numbers, then the birelation (I, R) is logically homogeneous. Now let R be logically inhomogeneous, say isomorphic to the chain of natural numbers followed by the chain of negative and nonnegative integers, so that R is logically equivalent to I. Then the selector transforming (I, R) into its second relation R is a logical operator, even a free operator, which transforms a logically homogeneous birelation into a logically inhomogeneous relation (this example is due to A. Roberty).

A treatment of logical homogeneity quite similar to ours may be found in [VAU, 1961 and 1963]. For any finite subset F of the base of R, every logical automorphism (see 1.3.3) of the domain F may be extended to an automorphism of R.

1.8. Rigid and logically rigid multirelations

Let R be a multirelation. An element a of the base $|R|$ is called a *fixed element* if it is invariant under all automorphisms of R. The multirelation R is said to be *rigid* if all elements of $|R|$ are fixed. For example, any ordinal is rigid.

Given any two natural numbers k, p, an element a of $|R|$ is called a *(k, p)-fixed element* if the only (k, p)-automorphism of R whose domain is the singleton $\{a\}$ is the identity mapping on a. If such numbers k and p exist, we call a a *logically fixed* element. An element a is logically fixed if and only if there exists a unary formula P with predicates replaceable by R such that $P(R)(a)=+$ and $P(R)(b)=-$ for all $b\neq a$ in $|R|$.

A multirelation R is said to be *logically rigid* if all elements of $|R|$ are logically fixed. It is obvious that any logically rigid multirelation is rigid. However, the converse is false: since the set of all formulas is denumerably infinite, any logically rigid multirelation necessarily has a finite or denumerable base; thus a nondenumerable ordinal is rigid but not logically rigid.

The following proposition is proved in [MON-VAU, 1959]:

Let R *be a multirelation; then the following two conditions are equivalent:*

(1) *The restriction of* R *to its logically fixed elements is a logical restriction of* R.

(2) *For any unary formula* P *with predicates replaceable by* R, *there exists a unary formula* P* *such that the following two bound formulas take the*

value + *for* R:

$$\underset{x}{\exists} P(x) \Rightarrow \underset{x}{\exists}(P(x) \bigwedge P^*(x)),$$

$$\underset{xy}{\forall}(P^*(x) \bigwedge P^*(y)) \Rightarrow x \equiv y.$$

Intuitively speaking, condition (2) states that any existence theorem concerning R may be strengthened to give a uniqueness theorem.

Note that *if the restriction* S *of* R *to its logically fixed elements is logically equivalent to* R *and* S *is logically rigid, then* S *is a logical restriction of* R. Indeed, since S is a logical restriction of itself, it satisfies condition (1), and hence also condition (2). But R is logically equivalent to S and therefore satisfies condition (2), hence also condition (1).

On the other hand, the restriction S of R to its logically fixed elements may be logically equivalent – even isomorphic – to R, and yet not a logical restriction of R. For example, let R be the usual ordering of the natural numbers, $0 \leqslant 1 \leqslant 2 \leqslant \cdots$, with elements added as follows: $1' \leqslant 1$ and $1'$ is incomparable with 0; $2' \leqslant 2$ and $2'$ is incomparable with 0, 1, $1'$; and so on: $i' \leqslant i$ and i' is incomparable with 0, 1, $1'$, 2, $2'$,..., $i-1$, $(i-1)'$. Then S is the restriction of R to all the elements except 0 and $1'$; it is isomorphic to R but not a logical restriction (M. Pouzet).

1.8.1. Anticipating Chapter 5 for the definition of a theory, let us call $\mathscr{T}$ a *unicitary theory* if any model R of $\mathscr{T}$ satisfies (1) or (2) above. It is proved in [MON-VAU, 1959] that a theory $\mathscr{T}$ is unicitary if and only if, for any unary formula P with the same predicates as $\mathscr{T}$, there is a formula P* such that the two bound formulas in (2) belong to $\mathscr{T}$ (independently of any model of $\mathscr{T}$).

EXERCISES

1

Let A, A′ be two chains and k, p two integers. By 1.2.7, A and A′ are (k, p)-equivalent if and only if, for any $q \leqslant p$ and any decomposition of the form

$$A = A_0 + 1 + A_1 + 1 + \cdots + 1 + A_q$$

there exists a decomposition

$$A' = A'_0 + 1 + A'_1 + 1 + \cdots + 1 + A'_q,$$

such that the chains A_i and A'_i are $(k-1, p-q)$-equivalent for $i = 0, 1, \ldots, q$, and the same holds with A and A′ interchanged. Prove the following assertions.

(1) For any nonzero ordinal a, the ordinals ω^k and $a\omega^k$ are $(2k, p)$-equivalent for any p. Moreover, if p is positive and b a nonzero ordinal, then $(p+1)\,\omega^k$ and $b(p+1)\,\omega^k$ are $(2k+1, p)$-equivalent. Hence deduce that if a is a nonzero ordinal, then ω^ω and $a\omega^\omega$ are logically equivalent.

(2) The ordinals ω^{k+1} and $\omega^{k+1} + \omega^k$ are $(2k+1, p)$-equivalent for any p.

(3) No ordinal smaller than ω^k is $(2k, 2k)$-equivalent to ω^k. Moreover, no ordinal $\leqslant \omega^k$ can be $(2k+1, 2k+1)$-equivalent to an ordinal $> \omega^k$. Hence deduce that the only ordinal $(2k+1, 2k+1)$-equivalent to ω^k is ω^k itself.

(4) The ordinals ω^k and ω^{k+1} are $(2k, p)$-equivalent for any p, but not $(2k+1, 2k+1)$-equivalent.

(5) The ordinals ω^{k+1} and $\omega^{k+1} + \omega^k$ are $(2k+1, p)$-equivalent for any p, but not $(2k+2, 2k+2)$-equivalent.

(6) Two distinct ordinals are logically equivalent if and only if they have the forms $a \cdot \omega^\omega + r$ and $b \cdot \omega^\omega + r$ ($r < \omega^\omega$; $a, b \neq 0$). Hence deduce that there are only denumerably many logical equivalence classes of ordinals.

2

(1) Let A and B be chains with disjoint bases E and F, respectively, A′ and B′ chains with disjoint bases E′ and F′. Consider the sums A + B and A′ + B′, defined on E∪F and E′∪F′, respectively, and prove by induction that if g is a (k, p)-isomorphism of A onto A′ and h a (k, p)-isomorphism of B onto B′, then the union $g \cup h$ is a (k, p)-isomorphism of A + B onto A′ + B′. Hence deduce the following theorem [BET, 1954]: If A is logically equivalent to A′ and B logically equivalent to B′, then A + B is logically equivalent to A′ + B′.

(2) Generalize the above exercise to the case of two sums $\sum A_i$ and $\sum A'_i$, each defined by starting from a (finite or infinite) chain I and replacing each element i by A_i or by a chain A'_i logically equivalent to A_i.

(3) Let A and B be chains, and define a product A(B) as the generalized sum obtained by replacing each element of A by a chain isomorphic to B. Let g be a (k, p)-isomorphism of A onto A′. With each pair $(a_i, a'_i) \in g$ (where a_i is in the base of A and a'_i in that of A′), associate a (k, p)-isomorphism h_i of B onto B′. Show by induction that the union of the mappings h_i is a (k, p)-isomorphism of A(B) onto A′(B′). Hence deduce that if A is logically equivalent to A′ and B to B′, then A(B) is logically equivalent to A′(B′) (see [FEF. 1955]).

(4) Generalizing the preceding alinea, consider two chains A and A′: replace each element i of the base of A by a chain B_i, to obtain a sum chain $\sum A(B_i)$; do the same for the elements j of the base of A′, to obtain the sum $\sum A'(B'_j)$. With each logical class $\mathscr{C}$ we associate a unary relation C defined on the elements i: $C(i) = +$ or $-$ according as $B_i \in \mathscr{C}$

or not; similarly, we define a relation C′ on the elements j. Given two integers k, p, let $C_1, \ldots, C_h$ be the relations of this type on the base of A associated with the different (k, p)-equivalence classes, and $C'_1, \ldots, C'_h$ the relations on the base of A′ associated with these classes. Let g be a (k, p)-isomorphism of $(A, C_1, \ldots, C_h)$ onto $(A', C'_1, \ldots, C'_h)$; for each i mapped by g onto $j = g(i)$, let h_i be a (k, p)-isomorphism of B_i onto B'_j. Show by induction that the union of the mappings h_i is a (k, p)-isomorphism of $\sum A(B_i)$ onto $\sum A'(B'_j)$. As a particular case, we obtain the result of part 2 above, where $A = A'$ and each relation C coincides with the corresponding C′; the result of part 3 is also obtained, with C a constant relation (either + or −) and C′ the same constant.

3

(1) Consider the binary succession relation C on the integers: $C(x, y) = +$ if $y = x + 1$. Show that the only restriction of C logically equivalent to C is C itself. Thus, if R is a unary relation over the integers, the only logically equivalent restriction of (C, R) is (C, R) itself.

(2) Suppose that for every finite sequence s of truth values + and − there exists an integer a such that $R(a), R(a+1), \ldots$ reproduces s and moreover $R(x) = +$ for all negative x. Let R′ be another relation satisfying the analogous condition with $R'(x) = -$ for all negative x. Note that no logically equivalent restriction of (C, R) and logically equivalent restriction of (C, R′) can be isomorphic.

(3) We now wish to show that (C, R) and (C, R′) are logically equivalent. Let f be a bijective mapping of a finite set F of integers onto another such set F′, and k, p two natural numbers. Let us suppose the elements of F partitioned into disjoint intervals, which we shall call R-*fragments*, and F′ similarly partitioned into R′-fragments corresponding in bijective fashion to the R-fragments. We assume that the following conditions (compare 1.1.7!) are satisfied. Two corresponding fragments are intervals of the same length; the translation mapping one onto the other takes each element a of f into the element $f(a)$ of F′ and each element not in F to an element not in F′. There are at least $(p+1)^k$ elements between any two R-fragments (excluding the endpoints of the fragments). Finally, the translation mapping an R-fragment onto the corresponding R′-fragment, extended to $(p+1)^k$ elements preceding the fragment and $(p+1)^k$ elements following it, is a local isomorphism of R onto R′, hence also of (C, R) onto (C, R′). Under these assumptions, *f is a (k, p)-isomorphism of* (C, R) *onto* (C, R′). Note that, on the one hand, for any k and p there exists a mapping f satisfying these conditions. On the other hand, our assertion is obvious for $k = 0$ and may be proved for $k \geqslant 1$ by induction, since when an interval containing $(p+1)^k$ elements is partitioned by at most p elements there remains at least one interval containing $(p+1)^{k-1}$ elements.

This example is due to D. Lacombe (1966; unpublished); another example of logically equivalent relations which have no common restriction logically equivalent to each was obtained previously by [KAR, 1963]; see Exercise 7.

(4) Using 1.7.2, deduce that the class of birelations logically equivalent to (C, R) contains no logically homogeneous birelations.

Problem. Does there exist a logical class not containing any logically homogeneous multirelations?

4

(1) Let I be the chain of natural numbers, C the succession relation on the natural numbers, and $\mathscr{A}$ the class of birelations logically equivalent to (I, C), which is at the same time a logical class (see 1.5.2). Consider the selector $\mathscr{P}$ which transforms each (2, 2)-ary birelation into its second relation: it is a free and therefore logical operator. Show that

the image $\mathscr{P}(\mathscr{A})$ is the class of relations logically equivalent to C; it is not a logical class, for C is not finitely-axiomatizable (1.5.3).

(2) Let I′ be the chain obtained by adding a chain isomorphic to the chain of integers after I. Let A′ be a unary relation taking the value + for exactly one of the integers following I. Let $\mathscr{A}$ be the class of all birelations logically equivalent to (I′, A′), and $\mathscr{P}$ the selector transforming each (2, 1)-ary birelation into its first relation. Recall that I′ is a logical extension of I; note that there is no relation A on the natural numbers such that (I, A) is logically equivalent to (I′, A′). Hence show that the image class $\mathscr{P}(\mathscr{A})$, which consists of relations logically equivalent to I, does not contain I itself and is therefore not a logical equivalence class. Hence deduce that the image of a logical class under a logical operator is not necessarily closed under logical equivalence.

(3) Even if $\mathscr{A}$ is both a logical class and a logical equivalence class, the image of $\mathscr{A}$ under a free operator need not be a logical equivalence class. (Consider the same chain I′ as above, and let A′ be the relation equal to + for the natural numbers and to − for all other integers.)

5

Let C be the succession relation on the integers: $C(x, y) = +$ if $y = x+1$. Let R be the semi-chain coinciding with the usual chain of natural numbers except that $R(x, y) = -$ whenever at least one of the integers x, y is negative; in other words, $R(x, y) = +$ if and only if $0 \leqslant x \leqslant y$. Let R′ be the symmetric semi-chain: $R'(x, y) = +$ if and only if $x \leqslant y \leqslant 0$. Let I denote the usual chain of integers and Z the unary relation taking the value + for the single element 0.

(1) Show that Z and I are interpretable by the birelation (R, R′), hence also by (C, R, R′), and that C, R and R′ are interpretable by (Z, I). Hence deduce that (C, R, R′) is finitely-axiomatizable (see Volume 1, 7.4.5).

(2) Given two natural numbers k and p, consider a set $a_1, \ldots, a_r$ of elements distinct from the integers, where $r = (p+1)^k$; define extensions R* and C* of R and C, respectively, by setting $R^*(x, y) = -$ whenever either x or y is one of the a_i $(i = 1, \ldots, r)$, and

$$C^*(a_1, a_2) = C^*(a_2, a_3) = \cdots = C^*(a_r, a_1) = +$$

(so that the elements a_i constitute a cycle), $C^*(x, y) = -$ in all other cases (if x is not followed by y in the cycle or if either x or y is an a_i and the other is an integer). Using 1.1.7, show that the birelations (C, R) and (C*, R*) are (k, p)-equivalent, hence deducing that (C, R) is not finitely-axiomatizable. Then, setting M = (C, R) and M′ = (C, R′), show that neither M nor M′ is finitely-axiomatizable, though their concatenation MM′ is finitely-axiomatizable (this example was suggested by J. P. Calais).

6

Let N denote the chain of natural numbers, Z the chain of integers, Q the chain of rational numbers. Recall that N and N+Z are logically equivalent (see 1.1.6 and 1.2.2).

(1) Show that the chains N+Q and N+Z+Q are logically equivalent (use Exercise 2), and that each is embeddable in the other, i.e., each is isomorphic to a restriction of the other.

(2) Show that if Z is an interval of a chain R, it remains an interval in any logical extension of R. Hence conclude that no logical extension of N+Z+Q is isomorphic to N+Q.

(3) Show that N+Q(Z) and N+Q(Z)+Z are not isomorphic, each of these relations has an initial interval isomorphic to the other and each of them is a logical extension of this initial interval. (Use 1.1.6. This example is due to [TAR-VAU, 1957], pp. 90–91.)

(4) Show that N+Q+N+Z and N+Z+Q+N are embeddable in one another, they are logically equivalent but neither of them is isomorphic to a logical extension of the other (this example is due to J. L. Paillet).

7

Let G denote the additive group of pairs (x, y), where x is an integer and y a dyadic rational number (i.e., with denominator a power of 2). Consider the pairs of rational numbers $a=(1,0)$, $b_0=(0,1)$, $b_1=(1/2)(1,1)$, $b_2=(1/8)(3,1)$, and for each positive integer n

$$b_n=(1/2^{1+2+\cdots+n})((1+2+2^{1+2}+\cdots+2^{1+2+\cdots+n-1}),1);$$

Let H denote the additive group of pairs of rational numbers generated by $a, b_0, b_1, \ldots, b_n, \ldots$.

Show that G and H are logically equivalent. However, there are no subgroups G′ and H′ of G and H, respectively, logically equivalent to G and H, such that G′ and H′ are isomorphic ([KAR, 1963], using [ERS, 1963]; communicated by J. Adda and M. H. Dulac).

8

(1) Let C be the succession relation on the set E of natural numbers. Let C′ be a relation isomorphic to C, defined on the same base E, and interpretable in C. Show that there exists a finite set F of natural numbers such that $C' \mid E-F = C \mid E-F$.

Hint. Let (k, p) be the characteristic of a formula interpreting C′ in C. By 1.1.7 with $r=(p+1)^k$, for all $i, j \geqslant r$ the mapping $f(i)=j, f(i+1)=j+1, \ldots$ is a (k, p)-automorphism of C and hence a local automorphism of C′. Let $j>i$ such that $C'(i,j)=+$, and set $u=j-i$; then $C'(i, i+u)=C'(i+u, i+2u)=\cdots=+$. By virtue of the isomorphism of C and C′, we must have $u=1$.

(2) Let C be the succession relation on the set E of all integers, C′ a relation isomorphic to C and interpretable in C. Show that there exists a finite set F such that $C' \mid E-F$ is either equal to $C \mid E-F$ or the symmetric image of the latter.

(3) In either of the above two cases, if C′ is isomorphic to C and interpretable in C, then C is interpretable in C′.

These results are due to A. Roberty (1970, unpublished).

9

Given integers k, p, call an operator $\mathscr{P}$ a *(k, p)-operator* if any (k, p)-isomorphism of any R onto any R′ is a local isomorphism of $\mathscr{P}(R)$ onto $\mathscr{P}(R')$ (see 1.3.4). Redefine the quantifier $\underset{1}{\forall_1^0}$ as a (1, 1)-operator which transforms any unary relation R into a 0-ary relation $\mathscr{P}(R)$ such that $\mathscr{P}(R) \vdash R$. Show that this new definition yields *two* universal quantifiers, which transform the unary relation on the empty base (see Volume 1, 3.6.2) into $(\emptyset, +)$ and $(\emptyset, -)$, respectively.

CHAPTER 2

LOGICAL CONVERGENCE; COMPACTNESS, OMISSION AND INTERPRETABILITY THEOREMS

2.1. Logical convergence

We now provide the class of multirelations of a given arity with a topology by stipulating that the open sets be all possible unions of logical classes. It is clear that any union of open sets is open and the intersection of any two open sets is open, so that this is indeed a topology. The closed sets, or intersections of logical classes, will be called *δ-logical classes.*

Every logical class is both open and closed, since its complement is a logical class; the converse will be proved in 2.2.5 below.

We shall say that a sequence of multirelations R_n, $n=1, 2, \ldots$, *converges to a multirelation* S if, for any logical class $\mathscr{A}$ containing S, there exists an integer m such that $R_n \in \mathscr{A}$ for all $n \geqslant m$. An equivalent definition is: for any k and p, there exists m such that, for all $n \geqslant m$, R_n is (k, p)-equivalent to S.

Given a class $\mathscr{A}$ of multirelations of the same arity, the *closure* of $\mathscr{A}$ is the smallest δ-logical class containing $\mathscr{A}$, i.e., the intersection of the logical classes containing $\mathscr{A}$. Equivalently, the closure is the class of all multirelations which are limits of convergent sequences of members of $\mathscr{A}$.

Any δ-logical class is closed with respect to logical equivalence. The converse, however, is false: the class of finite sets is closed with respect to logical equivalence but is not a δ-logical class (Volume 1, 5.6.3). Another example is the class of all succession relations with cycles (see 1.5.3).

2.1.1. Logical convergence lemma. *Let R_n, $n=1, 2, \ldots$, be a sequence of multirelations of the same arity, with bases E_n, respectively, and for each n let D_n be a subset of E_n satisfying the following conditions:*

(1) $D_n \subseteq D_{n+1}$.

(2) *The identity mapping on* D_n *is an* (n, n)*-isomorphism of* R_n *onto* R_{n+1}.

Then there exists a multirelation S, *defined on the union* D *of the subsets* D_n, *which is a common extension of all the multirelations* $R_n \mid D_n$.

(3) *Suppose, moreover, that for any n and any set* $\bar{D}_n$ *obtained by adding* $m \leqslant n$ *elements of* E_n *to* D_n, *there exists an* $(n-1, n-m)$*-isomorphism of*

R_n *onto* R_{n+1} *whose restriction to* D_n *is the identity and which maps* $\bar{D}_n$ *onto a subset of* D_{n+1}.

Then, for each n, the identity mapping on D_n *is an* (n, n)*-isomorphism of* R_n *onto* S.

▷ Since the identity mapping on D_n is a local isomorphism of R_n onto R_{n+1}, the restriction $R_{n+1} \mid D_{n+1}$ is an extension of $R_n \mid D_n$ for each n. Hence there exists a (unique) multirelation S with base D equal to the union of the subsets D_n which is a common extension of the multirelations $R_n \mid D_n$. The identity mapping on each D_n is a local isomorphism, hence a $(0, p)$-isomorphism of R_n onto S for any p.

Suppose that the identity mapping on D_1 is a $(1, 1)$-isomorphism of R_1 onto S, the identity mapping on D_2 is a $(2, 2)$-isomorphism of R_2 onto S,..., the identity mapping on D_{n-1} is an $(n-1, n-1)$-isomorphism of R_{n-1} onto S; lastly, suppose that for each $r=0, 1, 2,\dots$ the identity mapping on D_{n+r} is an $(n-1, n-1)$-isomorphism of R_{n+r} onto S. We claim that then, for each r, the identity mapping on D_{n+r} is an (n, n)-isomorphism of R_{n+r} onto S; the lemma will then follow by induction.

To prove our assertion, we let f_r denote the identity mapping on D_{n+r}; this is obviously both a $(0, n)$-isomorphism and an $(n-1, n-1)$-isomorphism of R_{n+r} onto S. Suppose that for some k $(1 \leqslant k \leqslant n)$ and all $r \geqslant 0$ the mapping f_r is a $(k-1, n)$-isomorphism of R_{n+r} onto S; we shall show that f_r *is a* (k, n)*-isomorphism of* R_{n+r} *onto* S.

Let $\bar{D}_{n+r}$ be obtained from D_{n+r} by adjoining $m \leqslant n$ elements of E_{n+r}. By condition 3 of our lemma, there exists a mapping $\bar{f}_r$ which is an $(n+r-1, n+r-m)$-isomorphism, hence also a $(k-1, n-m)$-isomorphism, of R_{n+r} onto R_{n+r+1}, coincides with the identity f_r on D_{n+r} and maps $\bar{D}_{n+r}$ onto a subset of D_{n+r+1}. By the induction hypothesis, the identity mapping on $\bar{f}_r(\bar{D}_{n+r})$ is a $(k-1, n)$-isomorphism, hence also a $(k-1, n-m)$-isomorphism, of R_{n+r+1} onto S. Thus $\bar{f}_r$ itself is a $(k-1, n-m)$-isomorphism of R_{n+r} onto S.

Conversely, let H be obtained from D_{n+r} by adjoining $m \leqslant n$ elements of D; then there exists $s \geqslant r$ such that $H \subseteq D_{n+s}$. By hypothesis, the identity mapping on H is a $(k-1, n)$-isomorphism, hence also a $(k-1, n-m)$-isomorphism, of S onto R_{n+s}. By condition 2, the identity mapping f_r on D_{n+r} is an $(n+r, n+r)$-isomorphism, hence also a (k, n)-isomorphism, of R_{n+s} onto R_{n+r}. Thus there exists an extension of f_r to H which is a

$(k-1, n-m)$-isomorphism of R_{n+s} onto R_{n+r} and is therefore also a $(k-1, n-m)$-isomorphism of S onto R_{n+r}.

The assertion is thus proved, and with it the entire lemma. ◁

2.2. Compactness theorem ([GOD, 1930]; see also [TAR, 1952], p. 710)

2.2.1. Assuming the axion of choice:

Let R_n be a sequence of multirelations of the same arity. Suppose that for any logical class $\mathscr{A}$ there exists m such that the R_n with $n \geqslant m$ are either all in $\mathscr{A}$ or all in the complement of $\mathscr{A}$. Then there exists a multirelation S with finite or denumerable base such that the sequence R_n converges to S.

▷ It will suffice to prove the existence of an infinite convergent subsequence R'_n of R_n. Moreover, we may replace each multirelation by an isomorphic image without affecting convergence of the sequence. Letting $\mathscr{A}$ be a (k, p)-equivalence class, we see that for any k and p there exists m such that the R_n with $n \geqslant m$ are pairwise (k, p)-equivalent. We define $R'_0 = R_0$ and associate with this multirelation the empty set D_0. Suppose that we have a subsequence of the given multirelations, $R'_1, \ldots, R'_n, \ldots, R'_{n+r}, \ldots$, each associated with a finite subset of its base: D_1 of cardinality $d(1)$ for $R'_1, \ldots, D_n$ of cardinality $d(n)$ for $R'_n, \ldots$. Suppose that conditions 1, 2, 3 of 2.1.1 are satisfied for $0, \ldots, n-1$, and that moreover the identity mapping on D_n is an (n, n)-isomorphism of R'_n onto R'_{n+r} for each $r \geqslant 0$.

Let $a_1, \ldots, a_{d(n)}$ denote the elements of D_n. Given any integer $m \leqslant n$ and any multirelation R whose base is a superset of D_n, let us say that sequences $b_1, \ldots, b_m$ and $b'_1, \ldots, b'_m$ are equivalent relative to R if $(R; a_1, \ldots, a_{d(n)}, b_1, \ldots, b_m)$ and the system obtained when each b_i is replaced by b'_i $(1 \leqslant i \leqslant m)$ are (n, n)-equivalent. The number of the resulting equivalence classes is finite and bounded by a number which depends not on R but only on the arity of R and on the numbers $d(n)$, m, n (see 1.2.1). Let $d(n+1)$ be the product of this bound by n.

By assumption, for sufficiently large k the multirelations R'_k are pairwise $(n+2, d(n+1)+n+1)$-equivalent. Confining attention to $R'_{n+1}, R'_{n+2}, \ldots$, let $s \geqslant n+1$ be the smallest index such that all R'_k with $k \geqslant s$ satisfy this condition, and set $R''_{n+1} = R'_s$. Let D_{n+1} be a subset of the base of R''_{n+1} containing a sequence $b_1, \ldots, b_m$ in each of the classes defined above with $R = R''_{n+1}$; we may always assume that the cardinality of D_{n+1} is $d(n+1)$.

Note that $D_{n+1} \supseteq D_n$, for each a_i $(1 \leqslant i \leqslant d(n))$ alone constitutes a sequence which is equivalent only to itself. For each multirelation $R''_{n+2}, R''_{n+3}, \ldots, R''_{n+r}, \ldots$ following R''_{n+1}, there exists a bijective mapping defined on D_{n+1} which is an $(n+1, n+1)$-isomorphism of R''_{n+1} onto R''_{n+r} $(r \geqslant 1)$. Identifying all elements with their images under this mapping, we see that conditions 1 and 2 carry over from the case $n-1$ to the case n, and the identity mapping on D_{n+1} becomes an $(n+1, n+1)$-isomorphism of R''_{n+1} onto each R''_{n+r}. As to condition 3, it also carries over. In fact, if $\bar{D}_n$ is obtained from D_n by adding m elements, there exist m elements $b_1, \ldots, b_m$ of the base of R''_{n+1} which yield an $(n-1, n-m)$-equivalent system. But then we can choose $b_1, \ldots, b_m$ in D_{n+1} which preserve the (n, n)-equivalence, hence also the $(n-1, n-m)$-equivalence.

By the axiom of choice, there exists an infinite sequence $D_0, D_1, \ldots, D_n, \ldots$ of subsets each of which satisfies the above conditions. We thus have the assumptions of 2.1.1, and since D is the union of the finite sets D_n it is either finite or denumerable. ◁

2.2.2. THEOREM. *Any sequence of multirelations of the same arity contains a subsequence which converges to a multirelation with finite or denumerable base* (using the axiom of choice).

▷ Since the set of logical classes is denumerable, we can enumerate them: $\mathscr{A}_0, \mathscr{A}_1, \ldots, \mathscr{A}_m, \ldots$. There are infinitely many indices n such that $R_n \in \mathscr{A}_0$ or infinitely many such that $R_n \in$ complement of $\mathscr{A}_0$. We take one of these infinite subsequences, retain its first term, R'_0 say, and then apply the same reasoning to $\mathscr{A}_1$, to $\mathscr{A}_2$, etc. The result is an infinite subsequence R'_n of R_n which satisfies the assumptions of 2.2.1. ◁

2.2.3. Another formulation of the compactness theorem:

For any set of logical or δ-logical classes with empty intersection, there exists a finite subset of classes with empty intersection.

▷ Let $\mathscr{A}_1, \ldots, \mathscr{A}_n, \ldots$ denote the given classes (there are at most denumerably many). We may assume that $\mathscr{A}_n \supseteq \mathscr{A}_{n+1}$ for each n, for if necessary $\mathscr{A}_n$ may be replaced by $\mathscr{A}_1 \cap \mathscr{A}_2 \cap \cdots \cap \mathscr{A}_n$. If $\mathscr{A}_n$ is not empty for any n, let $R_n \in \mathscr{A}_n$; then $R_i \in \mathscr{A}_n$ for all sufficiently large i. By 2.2.2, there exists a subsequence R'_n of R_n which converges to a multirelation S. Then for each $\mathscr{A}_n$ we have $S \in \mathscr{A}_n$; otherwise, S would be in the complement of $\mathscr{A}_n$, so that $R_i \in$ complement of $\mathscr{A}_n$ for sufficiently large i, contrary to our previous result. Finally, the intersection of the classes $\mathscr{A}_n$ is not empty, since it contains S. ◁

Another proof, more intuitive but calling for a generalization of the concepts of relation, operator and logical class, is indicated in Chapter 4, Exercise 5.

2.2.4. *Let $\mathscr{A}$ be a logical or δ-logical class containing multirelations with bases of arbitrarily large finite cardinalities. Then $\mathscr{A}$ contains a multirelation with infinite base.*

▷ Define $\mathscr{A}_n$ as $\mathscr{A}$ restricted to multirelations defined on at least n elements. Each $\mathscr{A}_n$ is a logical class or an intersection of logical classes, as the intersection of $\mathscr{A}$ with the logical class of multirelations defined on $\geqslant n$ elements (see Volume 1, 5.4). By hypothesis, no class $\mathscr{A}_n$ is empty and each is contained in its predecessors; thus, by 2.2.3, their intersection is not empty. But this intersection can contain only multirelations with infinite base. ◁

Corollary. The class of all finite chains, all finite groups, etc., is closed with respect to logical equivalence, but is not a δ-logical class (see Volume 1, 5.6.3).

2.2.5. *The logical classes are exactly those classes that are both open and closed in the topology defined in* 2.1.

▷ We have already seen (2.1) that each logical class is open and closed. Conversely, let $\mathscr{A}$ be an open and closed class. It is a union of logical classes, which we may always enumerate as $\mathscr{B}_1, \mathscr{B}_2, \ldots$, with $\mathscr{B}_n \subseteq \mathscr{B}_{n+1}$ for each n. The complement $\mathscr{A}'$ of $\mathscr{A}$ is also a union of logical classes $\mathscr{B}'_n$, with $\mathscr{B}'_n \subseteq \mathscr{B}'_{n+1}$ for each n. The union of the classes $\mathscr{B}_n \cup \mathscr{B}'_n$ contains every multirelation; hence, for sufficiently large n, these classes are constant. Thus for sufficiently large n we have $\mathscr{A} = \mathscr{B}_n$, so that $\mathscr{A}$ is a logical class. ◁

2.2.6. The topology we have been considering is uniform: for each characteristic (k, p), where k and p are natural numbers, one obtains a neighborhood by taking the class of pairs of (k, p)-equivalent relations. *Every logical operator is uniformly continuous*: let $\mathscr{P}$ be a logical operator represented by a formula of characteristic (k, p) and transforming two relations R, S into $\mathscr{P}$(R), $\mathscr{P}$(S). Given two integers k', p', we may ensure that $\mathscr{P}(\mathrm{R})$ and $\mathscr{P}(\mathrm{S})$ will be in the same neighborhood defined by (k', p')-equivalence by stipulating that R and S be in the same neighborhood relative to $(k+k', p+p')$-equivalence (see 1.3.5).

There exist uniformly continuous operators which are not logical. An example is an operator transforming each set E into a chain over E which is discrete and has both maximal and minimal element whether E is finite or infinite. Then, given k and p, we must find two sets E, E′ whose images under the operator are (k, p)-equivalent chains; to this end, it suffices to let E and E′ be $(1, (p+1)^k)$-equivalent, either of the same cardinality or both of cardinality at least $(p+1)^k$ (see 1.1.6).

Let $\mathscr{P}$ act as the identity operator except on the chain of natural numbers and its isomorphic images, in which it interchanges 0 and 1. $\mathscr{P}$ is continuous and, moreover, for any relation R and any isomorphism f on the base of R, we have $\mathscr{P}(f(\mathrm{R}))=f(\mathscr{P}(\mathrm{R}))$; nevertheless, $\mathscr{P}$ is not a logical operator. Note that the operator of transitive closure defined in 1.6.3, which commutes with every bijective mapping but is not logical, is not continuous: it maps the two birelations (C, R) and (C, R′) of Exercise 3 (3), Chapter 1, which are logically equivalent, onto two birelations which are no longer logically equivalent.

Problem. If $\mathscr{P}$ is a logical operator containing a predicate ρ such that $\mathscr{F}_{\rho}\mathscr{P}$ is continuous, is the latter also a logical operator (suggested by A. Roberty)?

2.3. Omission theorem

2.3.1. Let R be a multirelation, F a finite subset of the base $|\mathrm{R}|$. Call F *principal* if there exist two natural numbers k, p such that every (k, p)-automorphism of R with domain F is a logical automorphism (see 1.3.3).

Let F, G *be two finite subsets of the base of* R*; suppose that* F *is not principal. Then, for every natural number i, there exists an (i, i)-automorphism g of* R *with domain* G *such that no mapping of* F *into* $g(\mathrm{G})$ *is a logical automorphism of* R.

▷ We shall assume that F is a singleton $\{a\}$; the proof carries over immediately to the general case. Let $a_1, \ldots, a_h$ be the elements of G. If none of the mappings defined by taking a onto a_i $(i=1, \ldots, h)$ is a logical automorphism, we are done. Suppose, then, that a_r is the first term in the sequence $a_1, \ldots, a_h$ such that the mapping $a \rightarrow a_r$ is a logical automorphism.

The singleton $\{a_r\}$, like $\{a\}$, is not a principal subset: for every i, there exists an element which is (i, i)-equivalent to a_r but not logically equivalent. Let i be so large that a is not (i, i)-equivalent to any of $a_1, \ldots, a_{r-1}$. There exists a sequence $a'_1, \ldots, a'_h$ which is (i, i)-equivalent to $a_1, \ldots, a_h$ although a_r and a'_r are not logically equivalent. Then a and a'_1 are not logically equivalent, or even (i, i)-equivalent: otherwise a and a_1 would be logically equivalent; the same is true of a and $a'_2, \ldots, a$ and a'_{r-1}. Moreover, a and a'_r are not logically equivalent, for otherwise a_r and a'_r would be logically equivalent, contradicting our assumption.

Iterating the procedure, we map all the elements $a_1, \ldots, a_h$ onto $g(a_1), \ldots, g(a_h)$, and the mapping g is an (i, i)-automorphism of R. ◁

2.3.2. THEOREM. *Let* R *be a multirelation and* F *a nonprincipal finite subset of its base. There exists a denumerable multirelation* S, *logically equivalent to* R, *such that no mapping defined on* F *is a logical isomorphism of* R *onto* S (in other words, subsets logically equivalent to F are omitted in the base of S).

▷ Let us say that two elements x, y in the base $|R|$ are equivalent if the identity mapping on F, extended by mapping x onto y, is a local automorphism of R. Let G be the (finite) set obtained from F by adding a representative of each equivalence class. Using 2.3.1, we let b be a.(1, 1)-automorphism of R with domain G such that no mapping of F into $g(G)$ is a logical automorphism of R.

Set $F_1 = F$, $G_1 = G$, $g_1 = g$, $i(1) = 1$, $F_2 = g(G)$, and let $i(2)$ be an integer greater than $i(1) = 1$ such that no mapping of F into F_2 is an $(i(2), i(2))$-automorphism of R. Given any $j \leqslant i(2)$. we shall say that two sequences $x_1, \ldots, x_j$ and $y_1, \ldots, y_j$ of elements of $|R|$ are equivalent if the identity mapping on F_2, extended by mapping x_i onto y_i, $i = 1, \ldots, j$, is an $(i(2)-1, i(2)-j)$-automorphism of R. Let G_2 be the (finite) set obtained from F_2 by adding the elements of one sequence from each of the resulting equivalence classes. Again using 2.3.1, we let g_2 be an $(i(2), i(2))$-automorphism of R with domain G_2 such that no mapping of F into $g_2(G_2)$ is a logical automorphism of R.

Repeating the argument, we obtain a strictly increasing infinite sequence of natural numbers $i(1) = 1, i(2), \ldots, i(n), \ldots$, a set F_n, a superset G_n of F_n and a bijective mapping g_n with domain G_n for each n. Replacing R by an isomorphic multirelation R_n for each n, we constuct mappings $g_1, \ldots, g_n$

such that g_1 is the identity on $F=F_1, \ldots, g_n$ the identity on F_n. The following four conditions are now satisfied: (1) $F_n \subseteq F_{n+1}$ for each n; (2) the identity mapping on F_n is an $(i(n), i(n))$-isomorphism of R_n onto R_{n+1}; (3) for any $j \leqslant i(n)$ and set $\bar{F}_n$ obtained by enlarging F_n with j elements of $|R_n|$, there exists an $(i(n)-1, i(n)-j)$-isomorphism of R_n onto R_{n+1} defined on $\bar{F}_n$ with range in F_{n+1}, which coincides with the identity mapping on F_n; (4) no mapping of F into F_n is an $(i(n), i(n))$-isomorphism of $R=R_1$ onto R_n.

By conditions (1), (2), (3) and 2.1.1, the common extension S of the restrictions $R_n \mid F_n$ to the (denumerable) union of the sets F_n is $(i(n), i(n))$-equivalent to R_n, hence also to R, for each n: it is thus logically equivalent to R. On the other hand, suppose there exists a logical isomorphism f of $R=R_1$ onto S with domain F. Then there is a natural number m such that $f(F) \subseteq F_n$ for $n \geqslant m$. Thus the product of f and the identity mapping on F_n, which is an $(i(n), i(n))$-isomorphism of S onto R_n, is an $(i(n), i(n))$-isomorphism of R onto R_n, contradicting condition (4). ◁

2.4. Interpretability theorem [SVE, 1959′]

This theorem is a stronger version of that in Volume 1, 7.5.

Let R, S *be two multirelations with the same base. Then* S *is interpretable in* R *if and only if, for any* R′ *of the same arity as* R *and any* S′ *of the same arity as* S *such that* R′S′ *is logically equivalent to* RS, *any automorphism of* R′ *is an automorphism of* S′. *Moreover, it is sufficient to consider the case that* R′S′ *has finite or denumerable base.*

Suppose that S is interpretable in R, i.e., $S=\mathscr{P}(R)$, where $\mathscr{P}$ is a logical operator. Then the multirelations R′S′ such that $S'=\mathscr{P}(R')$ constitute a logical class. Thus, if R′S′ is logically equivalent to RS, it must also belong to this logical class, and it follows from Volume 1, 7.1.2, that any automorphism of R′ is an automorphism of S′.

The converse will be proved in 2.4.5 below.

2.4.1. With each natural number $i=0, 1, 2, \ldots$, we associate two sets F_i, G_i and a bijective mapping h_i of F_i onto G_i. Suppose that for each i the union $F_i \cup G_i$ is a subset of the intersection $F_{i+1} \cap G_{i+1}$ and, moreover, that the restriction of h_{i+1} to F_i coincides with h_i.

Under these assumptions, the union H of the sets F_i is equal to the union of the sets G_i and *there exists a (unique) permutation of* H *whose restriction to each* F_i *coincides with* h_i. This permutation is known as the *inductive limit* of the mappings h_i $(i=0, 1, 2,\ldots)$.

2.4.2. *Let* R, S *be multirelations with the same base* E, *and suppose that* S *is not interpretable in* R. *Then for each natural number i there exist multirelations* R_i *and* S_i *with the same base* E_i, *two finite subsets* F *and* G *of the intersection* $\bigcap_{i=1}^{\infty} E_i$, *and a bijective mapping h of* F *onto* G, *such that the following conditions hold:*

(1) R_iS_i *is isomorphic to* RS *for each i.*

(2) *The restictions* $R_i \mid F\cup G$ *coincide for all i; the same holds for* $S_i \mid F\cup G$.

(3) *h is an (i, i)-automorphism of* R_i *but not a local automorphism of* S_i.

▷ By hypothesis, since S is not interpretable in R, there exists for each i an (i, i)-automorphism h_i of R which is not a local automorphism of S (see 1.6). Let F_i be the domain and G_i the range of h_i; we may assume that the cardinality of these sets is the maximum arity of S (see Volume 1, 4.1.4). There are only finitely many possible cardinalities for the unions $F_i\cup G_i$; replacing the sequence $i=0, 1, 2,\ldots$ by a suitable infinite subsequence, we may therefore assume that these cardinalities are all equal. Similarly, there are only finitely many multirelations with given arity and finite base; so that, by taking a suitable isomorphism for each i, we may replace RS by a certain multirelation R_iS_i, identify all the sets F_i with a single set F, all the G_i with a single G, and all the mappings h_i with a single bijective mapping h; finally, all the restrictions $R_i \mid F\cup G$ and $S_i \mid F\cup G$, respectively, are identified, and the assertion is proved. ◁

2.4.3. Let R, S have the same base E, and suppose that S is not interpretable in R. Let p be a natural number. We define a *sequence of rank p* associated with R, S as follows. For each i, the sequence will contain a set E_i serving as the common base of two multirelations R_i, S_i, two finite subsets F_i, G_i of E_i and a bijective mapping h_i of F_i onto G_i, satisfying the following conditions:

(1) R_iS_i is isomorphic to RS for each i.

(2) $F_i\cup G_i$ is a subset of $F_{i+1}\cap G_{i+1}$ for each $i<p$; $F_i=F_p$ and $G_i=G_p$ for $i\geqslant p$.

(3) h_{i+1} is an extension of h_i for each $i<p$; $h_i=h_p$ for $i\geqslant p$.

(4) For each i, h_i is an (i, i)-automorphism of R_i, but h_0 is not a local automorphism of S_0; thus no h_i is a local automorphism of S_i.

(5) The identity mapping on $F_i \cup G_i$ is an (i, i)-isomorphism of R_iS_i onto $R_{i+1}S_{i+1}$ for each $i \leqslant p$, and a (p, p)-isomorphism for $i \geqslant p$; note that as a result,

$$R_iS_i \mid (F_i \cup G_i) = R_{i+1}S_{i+1} \mid (F_i \cup G_i).$$

(6) For each $i < p$, any $j \leqslant i$ and any set D obtained by adding j elements of E_i to $F_i \cup G_i$, there exists an $(i-1, i-j)$-isomorphism of R_iS_i onto $R_{i+1}S_{i+1}$, whose restriction to $F_i \cup G_i$ is the identity, which maps D onto a subset of $F_{i+1} \cap G_{i+1}$.

2.4.4. When $p=0$, conditions (1) through (6) reduce to conditions (1), (2), (3) of 2.4.2. Thus, for any multirelations R and S such that S is not intepretable in R, there exists a sequence of rank 0 associated with R, S.

Suppose that there exists a sequence of rank p; then there exists a sequence of rank $p+1$ with the same multirelations R_i, S_i, $i \leqslant p$.

▷ Given an integer $q \leqslant p$, we partition the sequences of q elements of E_p into equivalence classes, putting $a_1, \ldots, a_q$ and $a'_1, \ldots, a'_q$ into the same class if the identity mapping on $F_p \cup G_p$, extended by taking a_1 to $a'_1, \ldots, a_q$ to a'_q, is a $(p-1, p-q)$-automorphism of R_pS_p. The number of equivalence classes thus defined is finite (see 1.2.1). We take a representative of each class, and henceforth use the notation $a_1, \ldots, a_q$ only for terms of these representative sequences. By condition (5), for each $i \geqslant p+1$ the identity mapping on $F_p \cup G_p$ is a (p, p)-isomorphism of R_pS_p onto R_iS_i. Thus, for each representative sequence $a_1, \ldots, a_q$ of elements of E_p there exist a sequence $b_{i,1}, \ldots, b_{i,q}$ of elements of E_i and a $(p-1, p-q)$-isomorphism which is the identity on $F_p \cup G_p$ extended by taking a_1 to $b_{i,1}, \ldots, a_q$ to $b_{i,q}$. For each $i \geqslant p+1$, let D_i denote the union of $F_p \cup G_p$ and the preceding elements b_i. Replacing the sequence of indices $i \geqslant p+1$ by a suitable infinite subsequence, we may assume that all the sets D_i have the same cardinality, and renumber the indices from $p+1$ on. Replacing each R_iS_i $(i \geqslant p+1)$ by a suitable isomorphic multirelation, we maintain the subsisting conditions $F_i = F_p$, $G_i = G_p$, $h_i = h_p$ for $i \geqslant p$, and obtain $D_i = D_{p+1}$ for each $i \geqslant p+1$.

For each $i \geqslant p+1$, there exists k such that for all $j \geqslant k$ the mapping h_j, which is a (j, j)-automorphism of R_j by (4), has an extension which is an

(i, i)-automorphism of R_j with domain and range both containing D_j. Again replacing the sequence of indices $i \geqslant p+1$ by a suitable subsequence, we may ensure that the following condition hold. For each $i \geqslant p+1$, the new mapping h_i, which is an extension of the old one and hence also of h_p, has domain and range again denoted by F_i and G_i, respectively, such that $F_p \cup G_p \subseteq F_i \cap G_i$. The cardinalities of the new sets F_i and G_i again constitute a finite set, each of them being at most twice the cardinality of D_{p+1}. We may thus assume that the sets F_i $(i \geqslant p+1)$ all coincide, as do the sets G_i and the mappings h_i. We thus have conditions (1) through (4) of 2.4.3 with $p+1$ instead of p. Moreover, condition (6) will hold for $i=p$, thus being satisfied with $p+1$ instead of p. Finally, condition (5) may be ensured by again extracting a suitable subsequence of the indices $i \geqslant p+1$. Indeed, the latter indices may be divided into finitely many equivalence classes, if we declare i to be equivalent to j $(i, j \geqslant p+1)$ when the identity mapping on $F_{p+1} \cup G_{p+1}$ is a $(p+1, p+1)$-isomorphism of $R_i S_i$ onto $R_j S_j$. ◁

2.4.5. We can now complete the proof of the interpretability theorem.

▷ Suppose that S is not interpretable in R and construct a sequence of rank 0 as in 2.4.2. Then, iterating the procedure of 2.4.4, construct a sequence of multirelations $R_i S_i$ satisfying conditions (1) through (6) of 2.4.3, with no restriction on the number p; this is immediate, since the passage from p to $p+1$ leaves the multirelations $R_i S_i$, $i \leqslant p$, unchanged.

Let H denote the union of the sets F_i, which is also the union of the G_i. Let R′ be the extension of the multirelations $R_i \mid F_i$ to the base H and S′ the relation of the $S_i \mid F_i$ to H. Let h denote the permutation of H which is the inductive limit of the mappings h_i (see 2.4.1). Any finite restriction of h is a restriction of all h_i for sufficiently large i, hence a local automorphism of R_i and so also of R′; thus h is an automorphism of R′ (see Volume 1, 4.1.4). On the other hand, h is an extension of h_0, and so it is not an automorphism of S′. Finally, by 2.1.1, for each i the identity mapping on $F_i \cup G_i$ is an (i, i)-isomorphism of $R_i S_i$ onto R′S′. Consequently, R′S′ is (i, i)-equivalent to RS for any i, hence logically equivalent to RS. ◁

2.4.6. The following immediate corollary of the interpretability theorem is similar to the interpretability result of Volume 1, 7.5.

S *is interpretable in* R *if and only if, for any* R′, S′, S″ *such that* R′S′ *and* R′S″ *are logically equivalent to* RS, *we have* S′ = S″.

2.5. EVERY INJECTIVE LOGICAL OPERATOR IS INVERTIBLE

The aim of this section is to extend the result of Volume 1, 4.3.6, concerning free operators, to arbitrary logical operators; we shall in fact prove the result for logical *multi-operators* – finite sequences of logical operators of the same predicarity.

There is a great variety of injective multi-operators. Familiar examples are the connector $\neg_n^n$ which transforms every n-ary relation into its negation, the rank-changer $(\rho x^2 x^1)_2^2$, which transforms every binary relation into its converse; in either case, distinct relations are transformed into distinct relations.

Now consider an arbitrary logical operator, for example, $\mathscr{P} = \underset{1}{\forall_2^2}$, which transforms every binary relation R into $S = \mathscr{P}(R)$, where $S(x_1, x_2) = +$ if and only if $R(x_1, x_2) = +$ for every x_1 in the base. Then $\mathscr{P}$ is not injective, but the bi-operator that transforms every binary relation R into $(R, \mathscr{P}(R))$ is injective, though not free: a local isomorphism of R onto R′ need not be a local isomorphism of $(R, \mathscr{P}(R))$ onto $(R', \mathscr{P}(R'))$. Nevertheless, one might say that this bi-operator, though not free, has a free component – the identity operator. This is no longer the case for the bi-operator that transforms every binary relation R into the birelation $(R \wedge \mathscr{P}(R), R \wedge \neg \mathscr{P}(R))$; that this bi-operator is injective follows from the fact that $R = (R \wedge \mathscr{P}(R) \vee (R \wedge \neg \mathscr{P}(R))$.

2.5.1. Let $\mathscr{P}$ be a logical multi-operator, transforming each ρ-ary multirelation into a σ-ary multirelation, and R a ρ-ary multirelation.

If $\mathscr{P}$ *is injective on the class of multirelations logically equivalent to* R, *then there exists a logical multi-operator* $\mathscr{Q}$ *such that* $\mathscr{Q}\mathscr{P}(R) = R$. In other words, R *and* $\mathscr{P}(R)$ *are interpretable in one another.*

▷ Suppose that there is no logical multi-operator transforming $S = \mathscr{P}(R)$ into R; then R is not interpretable in S. By 2.4, there exist R′ and S′ such that R′S′ is logically equivalent to RS, and an automorphism f of S′ which is not an automorphism of R′. Since $S = \mathscr{P}(R)$, it follows from the logical equivalence of RS and R′S′ that $S' = \mathscr{P}(R')$, and so $fS' = \mathscr{P}(fR')$. Since $fS' = S'$ and $fR' \neq R'$, this contradicts the injectivity of $\mathscr{P}$. ◁

2.5.2. *Suppose that the logical multi-operator* $\mathscr{P}$ *is invertible on* R: *there exists a logical multi-operator* $\mathscr{Q}$ *such that* $\mathscr{Q}\mathscr{P}(\mathrm{R})=\mathrm{R}$. *Then:*

(1) $\mathscr{Q}\mathscr{P}(\mathrm{X})=\mathrm{X}$ *for every* X *logically equivalent to* R, *and* $\mathscr{P}\mathscr{Q}(\mathrm{Y})=\mathrm{Y}$ *for every* Y *logically equivalent to* $\mathscr{Q}(\mathrm{R})$.

(2) $\mathscr{P}$ *maps the class of multirelations logically equivalent to* R *bijectively onto the class of multirelations logically equivalent to* $\mathscr{P}(\mathrm{R})$, *and* $\mathscr{Q}$ *is the (bijective) inverse of* $\mathscr{P}$.

▷ (1) Let $\mathscr{Q}\mathscr{P}(\mathrm{R})=\mathrm{R}$. The class of multirelations X such that $\mathscr{Q}\mathscr{P}(\mathrm{X})=$ $=\mathrm{X}$ is a logical class; this may be verified by considering the intersection of the logical class $\{(\mathrm{X}, \mathscr{Q}\mathscr{P}(\mathrm{X}))\}$ and the logical class $\{(\mathrm{X}, \mathrm{X})\}$ (see Volume 1, 5.4.1 and 5.4.2). This logical class contains R and so contains any X logically equivalent to R. Similarly, the class of multirelations Y such that $\mathscr{P}\mathscr{Q}(\mathrm{Y})=\mathrm{Y}$ is a logical class which contains $\mathscr{P}(\mathrm{R})$ and therefore contains any Y logically equivalent to $\mathscr{P}(\mathrm{R})$.

(2) If $\mathrm{Y}\neq\mathrm{Y}'$ are both logically equivalent to $\mathscr{P}(\mathrm{R})$, then $\mathscr{Q}(\mathrm{Y})\neq\mathscr{Q}(\mathrm{Y}')$, since $\mathscr{P}\mathscr{Q}(\mathrm{Y})=\mathrm{Y}\neq\mathrm{Y}'=\mathscr{P}\mathscr{Q}(\mathrm{Y}')$. Thus the restriction of $\mathscr{Q}$ to multirelations logically equivalent to $\mathscr{P}(\mathrm{R})$ is injective. That $\mathscr{P}$ and $\mathscr{Q}$ are mutually inverse follows from part 1. ◁

Problem. Let $\mathscr{P}$ be a logical operator which is invertible on R. Does there always exist a logical class $\mathscr{A}$, containing R, such that $\mathscr{P}$ is injective on $\mathscr{A}$ and $\mathscr{P}(\mathscr{A})$ is a logical class?

2.5.3. *Let* $\mathscr{P}$ *be a logical multi-operator, of predicarity* ρ *and arity* σ, *which is injective on the class of all* ρ*-ary multirelations. Then there exists a logical multi-operator* $\mathscr{Q}$ *such that* $\mathscr{Q}\mathscr{P}(\mathrm{X})=\mathrm{X}$ *for every* ρ*-ary multirelation* X (A. Roberty, 1970; unpublished).

▷ We first show that the class of ρ-ary multirelations may be partitioned into finitely many disjoint logical classes $\mathscr{A}_i$ for each of which there exists a logical multi-operator $\mathscr{Q}_i$ such that $\mathscr{Q}_i\mathscr{P}(\mathrm{R})=\mathrm{R}$ for $\mathrm{R}\in\mathscr{A}_i$. By 2.5.1, for each ρ-ary multirelation R there exists a logical multi-operator $\mathscr{Q}_\mathrm{R}$ such that $\mathscr{Q}_\mathrm{R}\mathscr{P}(\mathrm{R})=\mathrm{R}$, and the class $\mathscr{A}_\mathrm{R}$ on which $\mathscr{Q}_\mathrm{R}\mathscr{P}(\mathrm{X})=\mathrm{X}$ is a logical class. We thus have a cover of the class of ρ-ary multirelations by logical classes $\mathscr{A}_\mathrm{R}$. By the compactness theorem (2.2.3), we can extract a finite subcover $\mathscr{A}_1,\ldots,\mathscr{A}_h$. Moreover, we may assume without loss of generality that the classes $\mathscr{A}_i$ are pairwise disjoint (otherwise, we need only consider their intersections).

For each of our logical classes $\mathscr{A}_i$ ($i=1,\ldots,h$), let $\mathscr{Q}_i$ be a logical multi-operator such that $\mathscr{Q}_i\mathscr{P}(\mathrm{X})=\mathrm{X}$ for each $\mathrm{X}\in\mathscr{A}_i$. Since $\mathscr{P}$ is injective, the classes $\mathscr{P}(\mathscr{A}_i)$ are pairwise disjoint. For each i, let $\mathscr{B}_i$ be the logical class of multirelations Y such that $\mathscr{P}\mathscr{Q}_i(\mathrm{Y})=\mathrm{Y}$. We claim that $\mathscr{B}_i\supseteq\mathscr{P}(\mathscr{A}_i)$. Indeed, if $\mathrm{Y}\in\mathscr{P}(\mathscr{A}_i)$, there exists $\mathrm{X}\in\mathscr{A}_i$ such that $\mathrm{Y}=\mathscr{P}(\mathrm{X})$; thus $\mathscr{Q}_i(\mathrm{Y})= =\mathscr{Q}_i\mathscr{P}(\mathrm{X})=\mathrm{X}$, and so $\mathscr{P}\mathscr{Q}_i(\mathrm{Y})=\mathscr{P}(\mathrm{X})=\mathrm{Y}$. Moreover, if $\mathrm{Y}\in\mathscr{B}_i$ and $\mathrm{Y}\in\mathscr{B}_j$ ($i\neq j$), then $\mathscr{Q}_i(\mathrm{Y})=\mathscr{Q}_j(\mathrm{Y})$; for otherwise we would have $\mathscr{P}\mathscr{Q}_i(\mathrm{Y})\neq\mathscr{P}\mathscr{Q}_j(\mathrm{Y})$ by the injectivity of $\mathscr{P}$, and so $\mathrm{Y}\neq\mathrm{Y}$, which is absurd.

It is now clear that, using suitable formulas, one can construct a logical multi-operator $\mathscr{Q}$ which (a) coincides with $\mathscr{Q}_i$ on each $\mathscr{B}_i$, $i=1,\ldots,h$; (b) transforms every multirelation which is not in any class $\mathscr{B}_i$ into the constant relation $-$. This is the required multi-operator. ◁

Corollary. The image of a logical (or δ-logical) class under an injective logical operator is a δ-logical class (compare Exercise 4, Chapter 1).

2.5.4. *For any logical multi-operator $\mathscr{P}$ the class of multirelations on which $\mathscr{P}$ is invertible is a union of logical classes.* This is obvious: if there exists $\mathscr{Q}$ such that $\mathscr{Q}\mathscr{P}(\mathrm{X})=\mathrm{X}$, then the latter equality defines a logical class.

The above class need not be a logical class (A. Roberty).

▷ Let $\mathscr{P}$ be the operator represented by

$$\rho xy \wedge x \not\equiv y \wedge \forall_z \rho zx \vee \rho yz.$$

$\mathscr{P}$ transforms every finite chain R into the associated succession relation $\mathscr{P}(\mathrm{R})$; conversely, any finite chain is interpretable in the succession relation. But the chain $\omega+\omega^-$ (the positive integers followed by the negative integers) is not interpretable in the associated succession relation. Indeed, let $a\geqslant(p+1)^k$ be a positive integer (in the component ω), and $b\leqslant-(p+1)^k$ a negative integer (in the component ω^-). The mapping interchanging a and b is a (k,p)-automorphism of the succession relation (see 1.1.7), but it is not order-preserving. Thus all finite chains belong to the class on which $\mathscr{P}$ is invertible, but their logical limit $\omega+\omega^-$ (see 1.1.6) does not. ◁

Problem. If R and S are each interpretable in the other (and of the same arity), does there exist an injective logical multi-operator $\mathscr{P}$ such that $\mathscr{P}(\mathrm{R})=\mathrm{S}$ (this would be a generalization of Volume 1, 4.3.7)?

The same question may be asked in a more general situation: assume that the arity σ of S is greater than the arity ρ of R, in the sense that there is an injective mapping f of the sequence of terms of ρ into that of σ with the property that $f(i) \geqslant i$.

2.5.5. *The image of a logical class under any injective logical multi-operator is a logical class.*

▷ Let $\mathscr{P}$ be a multi-operator of predicarity ρ and arity σ, and $\mathscr{Q}$ a multi-operator such that $\mathscr{Q}\mathscr{P}(\mathrm{X})=\mathrm{X}$ for every ρ-ary multirelation X (see 2.5.3). Then the image under $\mathscr{P}$ of the class of all ρ-ary multirelations is the class of σ-ary multirelations Y such that $\mathscr{P}\mathscr{Q}(\mathrm{Y})=\mathrm{Y}$, which is a logical class $\mathscr{U}$. Now let $\mathscr{A}$ be any ρ-ary logical class. Then $\mathscr{P}(\mathscr{A})$ is the intersection of $\mathscr{U}$ and the class of σ-ary multirelations Y such that $\mathscr{Q}(\mathrm{Y})\in\mathscr{A}$. But the latter, as the inverse image of $\mathscr{A}$ under $\mathscr{Q}$, is a logical class (Volume 1, 5.4.2). ◁

EXERCISE

1

Let i be a positive integer, C_i the binary cyclic relation on the integers $1, \ldots, i$, equal to $+$ for $(1, 2), (2, 3), \ldots, (i-1, i)$ and $(i, 1)$. Let P be a formula which associates with any binary relation R a relation P(R) equal to $+$ when there exists a t such that $R(x, t) = R(t, y)$.

(1) Show that the relations C_i are logically convergent to the succession relation C on the integers, and also to the succession relation C^* defined by infinitely many components isomorphic to C (the method of 1.1.7). Hence deduce that the birelations $(C_i, P(C_i))$ tend to $(C, P(C))$ and to $(C^*, P(C^*))$.

(2) Considering the odd numbers i, show that the class of all birelations (R, S), where R and S are isomorphic binary relations, is not a logical class. Considering C and C^*, show that this class is not even closed under logical equivalence. Considering the even numbers i, show that the class of all birelations (R, S), where R and S are logically equivalent binary relations, is not a logical class. Compare this with the case of identical R and S (Volume 1, 5.4.1).

(3) Do the same for the class of all birelations (R, S) such that $R \mid X$ is isomorphic to $S \mid X$ for every subset X of the base.

CHAPTER 3

ELIMINATION OF QUANTIFIERS

3.1. Absolute eliminant

Let M, N be two multirelations with the same base. We shall call N an *absolute eliminant* of M if every local automorphism of N with finite domain is a (k, p)-automorphism of M for all k, p.

Examples. The chain Q of rational numbers is an absolute eliminant of itself: any local automorphism on a finite set of rationals may be extended to an automorphism of Q and is therefore a (k, p)-automorphism of Q for all k and p.

Let $Q^* = 1 + Q + 1$ be the rational numbers in a closed interval; this is a chain with both minimal and maximal element. Let A denote the singleton of the minimal element, i.e., the unary relation taking the value + for the minimal element alone; similarly, let B denote the singleton of the maximal element. Then the trirelation (Q^*, A, B) is an absolute eliminant of Q^*. However, Q^* is not an absolute eliminant of itself, for the transformation mapping the minimal element onto any other element of the base is a local automorphism of Q^* but not a $(1, 1)$-automorphism.

3.1.1. *Let* M′ *and* N′ *be two multirelations such that* MN *and* M′N′ *are logically equivalent. If* N *is an absolute eliminant of* M, *then* N′ *is an absolute eliminant of* M′.

▷ Let f' be a local automorphism of N′ with finite domain F′ and range $f'(F')$. By 1.3.3 and the fact that MN and M′N′ are logically equivalent, there exists for all k, p a (k, p)-isomorphism h of MN onto M′N′ with range $F' \cup f'(F')$. Let $F = h^{-1}(F')$ and let f be the bijective mapping of F onto itself induced from f' by h^{-1}. It is easy to see that f is a local automorphism of N, and so f is a (k, p)-automorphism of M. It follows that $f' = hfh^{-1}$ is a (k, p)-automorphism of M′. ◁

3.1.2. *Let* N *be an absolute eliminant of* M. *For every logical formula* P *with n-ary predicates replaceable by* M, *there exists a free formula* Q *with predicates replaceable by* N, *of the same arity n, such that* Q(N) $(a_1,\ldots,a_n)=$ $=$P(M) $(a_1,\ldots,a_n)$ *for any elements* $a_1,\ldots,a_n$ *of the base.*

For example, let M be the chain of rationals in the closed interval [0, 1] and P the formula $\underset{y}{\forall}\rho xy \vee \underset{y}{\forall}\rho yx$, where M is substituted for ρ. The value of P is + for 0 and 1, and − for any other element. Let N be the trirelation MAB, where A is the singleton 0 and B the singleton 1, and let α and β be two predicates replaceable by A and B; then Q is the free formula $\alpha x \vee \beta x$.

▷ To prove the proposition, it will suffice to show that there exists a free operator transforming N into the relation P(M). Let (k, p) be the characteristic of P (see 1.3); by assumption, any local automorphism of N with finite domain is a (k, p)-automorphism of M, therefore, by 1.3.1, a local automorphism of P(M). Thus P(M) is freely interpretable in N (since every local automorphism of N defined on $\leqslant n$ elements is a local automorphism of P(M) and by Volume 1, 4.2, this is sufficient). By Volume 1, 4.3.5, there exists a free operator transforming N into P(M). ◁

3.2. (k, p)-ELIMINANT

Let M, N be two multirelations with the same base and k, p two natural numbers. We shall call N a *(k, p)-eliminant* of M if every local automorphism of N with finite domain is a (k, p)-automorphism of M.

Examples. Let I be the chain of natural numbers and k, p two natural numbers. For every i, we let U_i denote the unary relation equal to + for all numbers $\geqslant i$ and V_i the binary relation such that $V_i(a, b) = +$ for two numbers a, b if and only if $b - a \geqslant i$. Then proposition 1.1.6 becomes: *The multirelation* $(I, U_0, U_1, \ldots, U_{h-1}, V_1, V_2, \ldots, V_h)$ *is a (k, p)-eliminant of* I *if* $h \geqslant (p+1)^k$. This example extends to the case of a discrete chain I with minimal element but no maximal element (see 1.1.6), with the relations U and V defined as before.

3.2.1. *Let k, p be two natural numbers and* M, N, M′, N′ *four multirelations such that* MN *and* M′N′ *are $(k+1, r)$-equivalent for all r. If* N *is a (k, p)-eliminant of* M, *then* N′ *is a (k, p)-eliminant of* M′.

The proof is the same as in 2.1.1, with k, p fixed.

3.2.2. *Let k, p be two natural numbers and* N *a* (k, p)*-eliminant of* M. *For any logical formula* P *with predicates replaceable by* M, *of arity n and characteristic* (k, p), *there exists a free formula* Q *with predicates replaceable by* N, *of the same arity n, such that* $Q(N)(a_1, \ldots, a_n) = P(M)(a_1, \ldots, a_n)$ *for all elements* $a_1, \ldots, a_n$ *of the base.*

The proof is the same as in 2.1.2, with k, p fixed.

3.3. Elimination algorithms for the chain of rational numbers and the chain of natural numbers

3.3.1. Chain of rationals. We denote this chain (or any other dense chain with neither minimal nor maximal element) by $\leqslant$. Let P be a formula of arity n, with a binary predicate replaceable by $\leqslant$. We may assume that P is a prenex formula with p quantifiers. By 3.1 and 3.1.2, there exists a free formula Q of arity n with a binary predicate, such that

$$Q(\leqslant)(a_1, \ldots, a_n) = P(\leqslant)(a_1, \ldots, a_n)$$

for all rationals $a_1, \ldots, a_n$. We now present an algorithm for the construction of Q. Suppose the algorithm is known for $p-1$ quantifiers. To fix ideas, we let $P = \underset{n+1}{\forall} P'$, where P' is an $(n+1)$-ary formula with $p-1$ quantifiers, and let Q' be an $(n+1)$-ary free formula such that

$$Q'(\leqslant)(a_1, \ldots, a_{n+1}) = P'(\leqslant)(a_1, \ldots, a_{n+1})$$

for all rationals $a_1, \ldots, a_{n+1}$.

With every sequence of rationals $a_1, \ldots, a_n$ we associate a *reduced* sequence $b_1, \ldots, b_n$, consisting of all or some of the numbers $b = 1, \ldots, n$, exhibiting the same order relations as the a_i $(i = 1, \ldots, n)$. The set of n-tuples of rationals is thus partitioned into finitely many equivalence classes, each determined by the order of the terms and the various equality or inequality relations between them, i.e., by a reduced sequence. For each such class, hence also for each reduced sequence $s = b_1, \ldots, b_n$, there exists an n-ary free formula Q_s such that $Q_s(\leqslant)(a_1, \ldots, a_n) = +$ if and only if the reduced sequence of $a_1, \ldots, a_n$ is s (the algorithm for construction of Q_s is obvious).

With each reduced sequence $s = b_1, \ldots, b_n$ we associate a formula Q_s^*, which is equal to Q_s or to the operator $-$ according as

$$\underset{n+1}{\forall}\ Q'(\leqslant)(b_1,\ldots,b_n,x^{n+1})=+ \quad \text{or} \quad -.$$

A finite number of trials suffices to check which of the above values is obtained. Indeed, the value is $+$ if $Q'(\leqslant)(b_1,\ldots,b_n,c)=+$ for all $c=b_1,\ldots,b_n$ and for a number c in each interval $(-\infty,1),(1,2),\ldots,(n-1,n)$, $(n,+\infty)$ whose endpoints are b's; thus we need only $2n+1$ trials. The value is $-$ if one of the values of Q' in these trials is $-$. Let Q be the disjunction of the formulas Q_s^* for all reduced sequences s. Then Q is the required free formula, for it satisfies the conditions

$$Q(\leqslant)(a_1,\ldots,a_n)=\underset{n+1}{\forall}\ Q'(\leqslant)(a_1,\ldots,a_n,x^{n+1})=P(\leqslant)(a_1,\ldots,a_n).$$

A similar argument holds when $P=\underset{n+1}{\exists}\ P'$; we need only replace the phrase "for all $c=b_1,\ldots,b_n$ and for a number c in each interval" by "for at least one $c=b_1,\ldots,b_n$ or a number c in some interval".

3.3.2. Chain of natural numbers. We again denote the chain (or any other discrete chain with minimal element but no maximal element) by $\leqslant$. Let P be an n-ary formula with a binary predicate replaceable by $\leqslant$; we again assume that P is prenex with p quantifiers. By 3.2 (with $k=p$) and 3.2.2, if we set $h=(p+1)^p$, there exists a free formula Q with predicates replaceable by $W_p=(\leqslant, U_1,\ldots, U_{h-1}, V_1,\ldots, V_h)$ such that $Q(W_p)(a_1,\ldots,a_n)=P(\leqslant)(a_1,\ldots,a_n)$ for all natural numbers $a_1,\ldots,a_n$. Recall the definitions: for each $i=1,\ldots,h$, $U_i(a)=+$ for all $a\geqslant i$ and $V_i(a,b)=+$ for all a, b such that $b-a\geqslant i$. We present an algorithm for construction of Q. Suppose we already have an algorithm for $p-1$ quantifiers. To fix ideas, let $P=\underset{n+1}{\forall}\ P'$, where P' is an $(n+1)$-ary formula with $p-1$ quantifiers and let Q' be an $(n+1)$-ary free formula such that

$$Q'(W_{p-1})(a_1,\ldots,a_{n+1})=P'(\leqslant)(a_1,\ldots,a_{n+1})$$

for all natural numbers $a_1,\ldots,a_{n+1}$.

With each sequence of natural numbers $a_1,\ldots,a_n$ we associate a *reduced* sequence $b_1,\ldots,b_n$, obtained by replacing each a by the smallest possible b giving W_p the same value. In other words, the minimal b is equal to the minimal a, unless the minimal a is $\geqslant h=(p+1)^p$, in which case $b=h-1$. The difference between each b and the next term b' (in increasing order) is equal to the difference between the original term a (replaced by b) and the next term a', unless $a'-a>h$, in which case $b'-b=h$. It is clear that the maximal b is at most $h-1+(n-1)\,h=nh-1$. The set of n-tuples of natural numbers is thus partitioned into finitely many equivalence classes, each uniquely determined by a reduced sequence. Moreover, for each such class, and hence also for each reduced sequence $s=b_1,\ldots,b_n$, there exists a free n-ary formula Q_s such that $\mathrm{Q}_s(\mathrm{W}_p)\,(a_1,\ldots,a_n)=+$ if and only if the reduced sequence corresponding to $a_1,\ldots,a_n$ is s (the algorithm for construction of Q_s is obvious).

With each reduced sequence $s=b_1,\ldots,b_n$ we associate a formula Q_s^* which is either Q_s or the operator $-$, according as

$$\underset{n+1}{\forall}\ \mathrm{Q}'(\mathrm{W}_{p-1})\,(b_1,\ldots,b_n,x^{n+1})=+ \quad \text{or} \quad -.$$

This truth value is determinable in finitely many trials. Indeed, by 3.2 (with $p-1$ for both k and p) the value is $+$ if $\mathrm{Q}'(\mathrm{W}_{p-1})\,(b_1,\ldots,b_n,c)=+$ for every integer c between 0 and $\max(b_1,\ldots,b_n)+h'$, where $h'=p^{p-1}$, i.e., $c=0,1,\ldots,nh-1+h'$. The value is $-$ otherwise, i.e., when Q' takes the value $-$ at least once. Let Q be the disjunction of the formulas Q_s^*, where s ranges over all reduced sequences. Then Q is the required formula, satisfying the equalities

$$\begin{aligned}\mathrm{Q}(\mathrm{W}_p)\,(a_1,\ldots,a_n)&=\underset{n+1}{\forall}\ \mathrm{Q}'(\mathrm{W}_{p-1})\,(a_1,\ldots,a_n,x^{n+1})\\&=\mathrm{P}(\leqslant)\,(a_1,\ldots,a_n).\end{aligned}$$

A similar argument holds for $\mathrm{P}=\underset{n+1}{\exists}\ \mathrm{P}'$, with the phrase "for every integer c" replaced by "for at least one integer c".

3.4. Positive dense sum; elimination of quantifiers over the sum of rational or real numbers

A *sum* is a ternary functional relation which is associative, commutative, has a neutral element, is simplifiable and totally orderable. These terms will be defined immediately.

A ternary relation S is said to be *functional* if, for any a, b in its base, there is exactly one element c such that $S(a, b, c) = +$; we write $c = a + b$ (mod S).

A ternary functional relation S is said to be *associative* if $(a+b)+c = a+(b+c)$ (mod S) and *commutative* if $a+b=b+a$ (mod S), for all a, b, c in the base.

An element 0 of the base is *neutral* for S if $0+a=a+0=a$ (mod S) for any a.

A functional relation S is said to be *simplifiable* if $a+c=b+c$ (mod S) implies that $a=b$ for all a, b, c, and similarly $c+a=c+b$ implies $a=b$. If S is simplifiable, then for all a, b there exists at most one element c such that $a+c=b$; if this element exists, we write $c=b-a$ (mod S).

Finally, S is said to be *totally orderable* if, for all a, b, there exists either c such that $a+c=b$ or d such that $b+d=a$, and moreover $a+\cdots+a=0$ (any number of terms, any a) entails $a=0$.

A sum S is said to be *positive* if $a+b=0$ implies $a=b=0$ for all a, b. In that case, the condition "$a \leqslant b$ when there exists x such that $a+x=b$" defines a chain (total ordering), i.e., a reflexive, transitive and antisymmetric relation such that any two elements of the base are comparable. Indeed, transitivity follows from the associativity of S; antisymmetry is proved thus: if $a+x=b$ and $b+y=a$, then $a+(x+y)=a+0$, so that $x+y=0$ (as S is simplifiable) and finally $x=y=0$ (as S is positive). The chain thus defined is said to be *associated* with the sum S. The neutral element 0 is the minimal element of the chain. If the base contains elements other than 0, there is no maximal element: any maximal element u would be such that $u+x=u$ for all x, so that $x=0$. The chain associated with S is *compatible* with S in the sense that $a \leqslant b$ implies $a+c \leqslant b+c$ and conversely, for all a, b, c. As usual, we write $a<b$ if $a \leqslant b$ and $a \neq b$.

Let h be a natural number. A sum S is said to be *h-dense* if, for any a, there exists b such that $b+\cdots+b$ (h times) $=a$. The element b is unique – this follows from the fact that S is totally orderable and positive. In particular, any two elements a, b with $a<b$ have a half-sum c, defined by $c+c=a+b$, such that $a<c<b$. Indeed, if $c \leqslant a$, say, we have $c+c \leqslant a+c<a+b=c+c$, which is absurd.

A sum is said to be *dense* if it is h-dense for any natural number h. An example of a dense sum is the usual sum of rational numbers. A sum which is 2-dense but not 3-dense is the usual sum of dyadic rationals (rational numbers whose denominators are powers of 2).

3.4.1. Let S denote the usual sum over the set E of nonnegative rational numbers, and S′ another positive dense sum with base E′. Let n be a natural number, a_i $(i=1,\ldots,n)$ a sequence of elements of E, and a'_i a sequence of elements of E′. Let p be a positive integer. We define the *subfactorial* of p (denoted $\operatorname{fac}(p)$) to be the smallest integer divisible by every natural number $\leqslant p$. Define a function u on the natural numbers by $u(0)=2$, $u(1)=8,\ldots, u(p+1)=2u(p)\operatorname{fac}(u(p))$.

A sufficient condition for the mapping that takes each a_i onto a'_i $(i=1,\ldots,n)$ to be a (p, p)-isomorphism of S *onto* S′ *is that the elements a_i and a'_i (separately) satisfy the same equalities and inequalities (*mod S *or* mod S′*, respectively) of the form*

$$u_1a_1+\cdots+u_na_n=0,\qquad >0,\qquad <0,$$

where the coefficients $u_1,\ldots,u_n$ range over all integers of absolute value $\leqslant u(p)$.

The meaning of these equalities and inequalities should be clear: transferring the terms with negative coefficients to the right and then replacing terms ua by $a+\cdots+a$ (u times), one obtains sums mod S (or mod S′) and inequalities relative to the chain associated with S (or S′) as defined in 3.4 above.

▷ The assertion is true for $p=0$, $u(0)=2$. Indeed, the mapping taking each a_i to a'_i will be a local isomorphism of S onto S′ if all equalities and inequalities of the types $a_i+a_j-a_k=0$ $(>0, <0)$, $2a_i-a_j=0$ $(>0, <0)$, $a_i=0$ (>0) $(i, j, k=1,\ldots,n)$ are preserved when S is replaced by S′ and each a by the corresponding a'.

Supposing the assertion true for p, we now prove it for $p+1$. The mapping $a_i\to a'_i$ $(i=1,\ldots,n)$ will be a $(p+1, p+1)$-isomorphism of S onto S′ if, for any $b\in$ E, there exists $b'\in$ E′ such that the mapping $a_i\to a'_i$, $b\to b'$ is a (p, p)-isomorphism, and conversely with a_i and a'_i interchanged (to obtain the case of several elements b instead of one, use the fact that any (q, q)-isomorphism is also a (p, q)-isomorphism for any $q\leqslant p$; see 1.1.1). Thus it will suffice to show that for any b there exists b' such that the elements a_i, b, on the one hand, and the elements a'_i, b, on the other, satisfy the same equalities and inequalities of the type

$$u_1a_1+\cdots+u_na_n+vb=0,\qquad >0,\qquad <0,$$

where the coefficients u and v are integers of absolute value $\leqslant u(p)$, and conversely with the elements a_i and a'_i interchanged.

Multiply each equality (inequality) by an integer such that the new coefficient of b (if not zero) is $\operatorname{fac}(u(p))$. The new coefficients of the a_i will be integers of absolute value $\leqslant u(p) \operatorname{fac}(u(p))$. The new equalities and inequalities are equivalent to the old ones, thanks to associativity and commutativity. The coefficients in the relations not involving b are of absolute value $\leqslant u(p)$, hence also $\leqslant u(p+1)$. By the induction hypothesis, they are preserved when each a_i is replaced by a'_i. The other relations may be expressed as the sum of a term $(\operatorname{fac}(u(p)))\, b$ and a second term $u_1 a_1 + \cdots + u_n a_n$ (we shall call this an a-sum), where each u_i is of absolute value $\leqslant u(p) \operatorname{fac}(u(p))$. When we replace each a_i by a'_i, we can always find an element b' satisfying the same equalities and inequalities mod S′ (this follows from the density of S′), provided that the a'-sums are arranged by S′ in the same order as are the a-sums by S. This will hold if, whatever the difference between two a-sums, the difference between the corresponding a'-sums has the same sign (or else both differences vanish). Now the coefficients in these differences are integers of absolute value not exceeding

$$2u(p) \operatorname{fac}(u(p)) = u(p+1).$$

By hypothesis, the resulting equalities (inequalities) are preserved when each a_i is replaced by a'_i. The same argument is valid with S and S′, a and a' interchanged. Note, by the way, that each of the equalities (inequalities) in question must be reduced to a form involving only positive coefficients. When one goes over to differences as above, one must therefore transfer certain terms from one side to the other, thus using the fact that the sums in question are simplifiable. ◁

3.4.2. *Any two positive dense sums with infinite bases are logically equivalent. Moreover, if one of them,* S′ *say, is an extension of the other* (S), *then* S′ *is a logical extension of* S.

▷ We first assume that S′ is an extension of S. For any finite subset F of the base of S, consider a sequence a_i $(i=1,\ldots,n)$ exhausting F. The conditions of 3.4.1 are satisfied with $a'_i = a_i$ $(i=1,\ldots,n)$ and any natural number p. Thus the identity mapping on F is a (p,p)-isomorphism of S onto S′ for any p, so that S′ is indeed a logical extension of S.

Given an arbitrary positive dense sum S′ with infinite base, take some element $a \neq 0$ and let S be the restriction of S′ to the multiples of a by all rational numbers u/v (where u and v are natural numbers, $v \neq 0$). The base of S is thus the set of all b such that

$$\underbrace{b + \cdots + b}_{v \text{ times}} = \underbrace{a + \cdots + a}_{u \text{ times}}.$$

We see that S is isomorphic to the sum over the rational numbers. By the preceding part of the proof, any positive dense sum is logically equivalent to the sum S of rationals. ◁

3.4.3. *The sum of nonnegative rational numbers is not finitely-axiomatizable.*

▷ For each prime p, let $\mathscr{A}_p$ denote the logical class of positive q-dense sums for all primes $q \leqslant p$. If p' is the first prime $> p$, then $\mathscr{A}_{p'}$ is a proper subclass of $\mathscr{A}_p$. Indeed, the sum of rationals with reduced denominator divisible only by primes $\leqslant p$ is in $\mathscr{A}_p$ but not in $\mathscr{A}_{p'}$. By 3.4.2, the intersection $\mathscr{A}$ of all classes $\mathscr{A}_p$ is the class containing the sum of nonnegative rationals and all logically equivalent relations. By the compactness theorem, $\mathscr{A}$ is not a logical class. ◁

3.4.4. The foregoing arguments carry over *mutatis mutandis* (with similar results) to the case that S and S′ are totally orderable dense commutative groups. We recall that a functional commutative ternary relation S is a *commutative group* if it has a neutral element 0 and for each a there exists an element b such that $a + b = 0$ (mod S) (b is known as the *inverse* of a). The group is said to be *totally orderable* if there exists a chain I over the same base as S which is *compatible* with S, in the sense that $a \leqslant b$ (mod I) implies $a + c \leqslant b + c$ (mod I) for all a, b, c. Density is defined as in 3.4 above. A positive dense sum is obtained by restricting S to the elements $\geqslant 0$ (mod I).

Any two totally orderable dense commutative groups S *and* S′ *are logically equivalent. If* S′ *is an extension of* S, *then it is a logical extension of* S.

The same is true if each group is replaced by a birelation (S, I), where I is a chain compatible with S.

3.5. Positive discrete divisible sum; elimination of quantifiers over the sum of natural numbers

Let S be a positive sum (see 3.4; the operation will be denoted by $+$). S is said to be *discrete* if there exists an element 1 which is the immediate successor of the neutral (minimal) element 0 in the chain associated with S. It follows that each element a has a successor, viz. $a+1$ (mod S), and each element $a\neq 0$ has a predecessor, viz. $a-1$ (see 3.4 for the operation $-$). We let 2, 3,... denote the elements $1+1$, $(1+1)+1$,.... An element a is a *multiple* of a positive integer p if there exists b such that $a=b+\cdots+b$ (p times).

Given a positive integer p, a positive discrete sum is said to be *p-divisible* if for any a there exists b such that $a=\underbrace{b+\cdots+b}_{p\text{ times}}+h$, where h is one of the numbers 0, 1,..., $p-1$. The element b is uniquely determined by a and p. The sum is said to be *divisible* if it is p-divisible for every positive integer p.

3.5.1. Let S denote the usual sum over the set E of natural numbers and S′ another positive discrete divisible sum with base E′. Let n be a natural number, a_i ($i=1,\ldots,n$) a sequence of elements of E and a'_i a sequence of elements of E′. We define the subfactorial fac(p) as before, as the smallest positive integer divisible by every natural number $\leqslant p$. Define a function u on E by $u(0)=2$, $u(1)=36,\ldots$, $u(p+1)=3\,\text{fac}(u(p)\,\text{fac}(u(p)))$.

A sufficient condition for the mapping taking each a_i onto a'_i to be a (p,p)-isomorphism of S *onto* S′ *is that the a_i and a'_i satisfy the same equalities, inequalities and multiplicative conditions of the form*

$$u_1a_1+\cdots+u_na_n+v=0,\qquad >0,\qquad <0;$$
$$(u_1a_1+\cdots+u_na_n+v)\ \text{mult}\ r;$$

where the coefficients u, v are integers of absolute value $\leqslant u(p)$ and r ranges over all positive integers not exceeding $u(p)$ (the notation b mult r is an abbreviation for "b is a multiple of r").

▷ The assertion is true for $p=0$, $u(0)=2$, for the equalities and inequalities that must be satisfied for the mapping in question to be a local isomorphism of S onto S′ are the same as in 3.4.1.

Supposing the assertion true for p, we now prove it for $p+1$. The mapping taking each a_i onto a'_i will be a $(p+1, p+1)$-isomorphism if, for any $b \in \mathrm{E}$, there exists $b' \in \mathrm{E}'$ such that the mapping $a_i \to a'_i$, $b \to b'$ is a (p, p)-isomorphism, and conversely with a_i and a'_i interchanged. Thus it will suffice to show that for every b there exists an element b' satisfying the same equalities, inequalities and multiplicative conditions of the form

$$u_1 a_1 + \cdots + u_n a_n + tb + v = 0, \qquad > 0, \qquad < 0;$$
$$(u_1 a_1 + \cdots + u_n a_n + v + (\operatorname{fac} u(p))\, b) \text{ mult } r,$$

where u, v, t are all integers of absolute value $\leqslant u(p)$ and r ranges over all positive integers $\leqslant u(p)$; and conversely, with a_i and a'_i interchanged.

Retaining the conditions in which the coefficient of b is zero, multiply each of the other conditions by an integer in such a way that the new coefficient of b is $\operatorname{fac}(u(p))$. Note that the new conditions are equivalent to the old ones, in view of the characteristic properties of positive discrete divisible sums. We write the new conditions as follows:

$$u_1 a_1 + \cdots + u_n a_n + v + (\operatorname{fac} u(p))\, b = 0, \qquad > 0, \qquad < 0;$$
$$(u_1 a_1 + \cdots + u_n a_n + v + (\operatorname{fac} u(p))\, b) \text{ mult } r,$$

where the integers u, v, r are of absolute value $\leqslant u(p) \operatorname{fac}(u(p))$.

(1) Suppose first that these conditions include at least one equality, say

$$(\operatorname{fac} u(p))\, b = -u_1 a_1 - \cdots - u_n a_n - v.$$

We then add as a new condition the inequality or equality

$$-u_1 a_1 - \cdots - u_n a_n - v > 0 \quad \text{or} \quad = 0$$

and the multiplicative condition

$$(u_1 a_1 + \cdots + u_n a_n + v) \text{ mult } \operatorname{fac} u(p).$$

The last two conditions involve only a's and not b; they are preserved when each a_j is replaced by a'_i, and the coefficients u, v and the integer $\operatorname{fac}(u(p))$ are bounded in absolute value by $u(p) \operatorname{fac}(u(p)) < u(p+1)$. Thus there exists $b' \in \mathrm{E}'$ such that $(\operatorname{fac}(u(p)))\, b' = -u_1 a'_1 - \cdots - u_n a'_n - v$.

Replace $(\operatorname{fac}(u(p)))\, b$ by $-u_1 a_1 - \cdots - u_n a_n - v$ on the left of all the equalities, inequalities and multiplicative conditions involving b. This gives modified conditions which now involve only a's, with coefficients of absolute value bounded by $2u(p)\operatorname{fac}(u(p))<u(p+1)$. These conditions are preserved when each a_i is replaced by a_i'; thus, replacing each expression of type $-u_1a_1'-\cdots-u_na_n'-v$ by $(\operatorname{fac}(u(p)))\,b'$, we recover the original conditions with a_i and b replaced by a_i' and b', respectively.

(2) Suppose that at least one of the expressions on the left of an inequality is of absolute value $\leqslant m=\operatorname{fac}(u(p)\operatorname{fac}(u(p)))$; for example,

$$0<u_1a_1+\cdots+u_na_n+v+(\operatorname{fac}u(p))\,b\leqslant m$$

(the reasoning for the case of an expression bounded by $-m$ and 0 is similar). Let w denote the (integer) value of the expression in question, so that $0<w\leqslant m$ and

$$(\operatorname{fac}u(p))\,b=-u_1a_1-\cdots-u_na_n-v+w.$$

We now add as a new condition the inequality or equality

$$-u_1a_1-\cdots-u_na_n-v+w>0 \quad \text{or} \quad =0$$

and the multiplicative condition

$$(-u_1a_1-\cdots-u_na_n-v+w)\text{ mult }\operatorname{fac}u(p).$$

The last two conditions are preserved when each a_i is replaced by a_i'; the coefficients u, $-v+w$ and the integer $\operatorname{fac}(u(p))$ are bounded in absolute value by $2m<u(p+1)$. The argument now continues just as in part 1, the coefficients being bounded in absolute value by $3m=u(p+1)$.

(3) Now suppose that all the expressions on the left of our inequalities are of absolute value $>m=\operatorname{fac}(u(p)\operatorname{fac}(u(p)))$. Consider one of these expressions, of minimum positive value:

$$u_1^+a_1+\cdots+u_n^+a_n+v^++(\operatorname{fac}u(p))\,b>m$$

and another expression, which is negative and of minimum absolute value:

$$u_1^-a_1+\cdots+u_n^-a_n+v^-+\operatorname{fac}u(p))\,b<-m.$$

Thus $(\operatorname{fac}(u(p)))\,b$, which is nonnegative, is bounded above by

$$-u_1^-a_1-\cdots-u_n^-a_n-v^->m$$

and below by

$$-u_1^+ a_1 - \cdots - u_n^+ a_n - v^+ ;$$

the difference between the upper and lower bounds is $>2m$. If the lower bound is nonnegative, we retain it; if it is negative, we replace it by 0: the difference between the upper bound and the new lower bound is $>m$ in either case.

Consider all the multiplicative conditions in which the new coefficient of b is $\mathrm{fac}(u(p))$; the corresponding integers r are all $\leqslant u(p)\ \mathrm{fac}(u(p))$, and therefore possess a common multiple $m = \mathrm{fac}(u(p)\ \mathrm{fac}(u(p)))$. By virtue of the divisibility of S, there exist $c \in \mathrm{E}$ and an integer h $(0 \leqslant h < m)$ such that

$$(\mathrm{fac}\, u(p))\, b = \underbrace{c + \cdots + c}_{m \text{ times}} + h .$$

Since m is a multiple of $\mathrm{fac}(u(p))$, it is also true for h. Now the difference between $(\mathrm{fac}(u(p)))\, b$ and h is a multiple of m, hence of each of the numbers r figuring in the multiplicative conditions; thus we may replace each of the latter by the equivalent condition

$$(u_1 a_1 + \cdots + u_n a_n + v + h)\ \mathrm{mult}\, r .$$

These new multiplicative conditions involve only a's and not b. They are preserved when each a_i is replaced by a_i', since the coefficients u, $v+h$ and the integers r are bounded in absolute value by $2m < u(p+1)$. Even more: they will be preserved if we replace h by an element differing from h by a multiple of m. Now, by virtue of the divisibility of S′, there is an element differing from h by a multiple of m in any interval of m consecutive elements. In particular, consider the image of the upper bound

$$-u_1^- a_1' - \cdots - u_n^- a_n' - v^- ,$$

and that of the lower bound, which is

$$-u_1^+ a_1' - \cdots - u_n^+ a_n' - v^+ \quad \text{or} \quad 0,$$

as the case may be. The difference between these two numbers is again $>m$, like the difference between the bounds themselves. In fact, the coefficients of the inequality which states that the upper bound is $>m$ are of the type u^- and $m + v^-$; the inequality stating that the upper bound minus the lower bound is $>m$ has coefficients of the type $u^+ - u^-$ and $v^+ - v^- - m$, which are bounded by $3m = u(p+1)$.

Take some element of E′, between the images of the (nonnegative) lower bound and the upper bound, which differs from h by a multiple of m. This element is a multiple of $\mathrm{fac}(u(p))$, say $(\mathrm{fac}(u(p)))\,b'$, where $b' \in \mathrm{E}'$. We thus obtain the new multiplicative conditions by replacing each a_i and b by a_i' and b', respectively, or, equivalently, replacing h by $(\mathrm{fac}(u(p)))\,b'$ in the multiplicative conditions just derived. Moreover, we obtain the two inequalities

$$u_1^+ a_1' + \cdots + u_n^+ a_n' + v^+ + (\mathrm{fac}\, u(p))\, b' > 0,$$
$$u_1^- a_1' + \cdots + u_n^- a_n' + v^- + (\mathrm{fac}\, u(p))\, b' < 0.$$

The other inequalities are still satisfied, since the difference between any two left-hand sides involves only a's, with coefficients of absolute value bounded by $2u(p)\,\mathrm{fac}(u(p)) < u(p+1)$: the sign of these differences is thus preserved when each a_i is replaced by a_i'.

Finally, the above argument remains valid in the case that all the left-hand sides of the inequalities are positive (or all are negative). ◁

3.5.2. *Any two positive discrete divisible sums* S′ *and* S *are logically equivalent. Moreover, if* S′ *is an extension of* S, *it is a logical extension of* S.

▷ Suppose first that S′ is an extension of S. For any finite subset F of |S|, a sequence a_i $(i = 1, \ldots, n)$ exhausting F satisfies the conditions of 3.5.1 with $a_i' = a_i$ for every positive integer p. Thus the identity mapping on F is a (p, p)-isomorphism of S onto S′ for every p, and so S′ is a logical extension of S.

Given S′, let S be the restriction of S′ to the natural numbers, i.e., to the elements 0, 1 and the sums generated by 1. It is easy to see that S is isomorphic to the usual sum of natural numbers. Thus S′ is a logical extension of S, and so any positive discrete divisible sum is logically equivalent to the sum of natural numbers. ◁

The above discussion carries over with obvious modifications to the case that S and S′ are commutative groups, each totally orderable by a discrete chain with neither minimal nor maximal element, and divisible (in the same sense as defined in 3.5).

3.5.3. *The sum of natural numbers is not finitely-axiomatizable.*

▷ For each prime p, let $\mathscr{A}_p$ denote the logical class of positive discrete sums which are q-divisible for each prime $q \leqslant p$. If p' is the next prime after

p, then $\mathscr{A}_{p'}$ is a proper subclass of $\mathscr{A}_p$. Indeed, let S_p be the sum of pairs (a, b), where a is a rational number whose denominator is divisible only by primes $\leqslant p$ and b is an integer (a natural number if $a=0$, an arbitrary integer if $a>0$). This sum S_p is in $\mathscr{A}_p$ but not in $\mathscr{A}_{p'}$, for the pair $(1, 0)$ is not a multiple of p', of the type $(a, 0)+\cdots+(a, 0)$ (p' times) unless $a=1/p'$. By 3.5.2, the intersection $\mathscr{A}$ of the classes $\mathscr{A}_p$ is the class containing the sum of natural numbers and all logically equivalent relations. Now $\mathscr{A}$ is not a logical class, for otherwise the differences $\mathscr{A}_p-\mathscr{A}$ would be logical classes with empty intersection; by the compactness theorem (2.2.3), this would imply that $\mathscr{A}_p=\mathscr{A}$ for sufficiently large p. ◁

3.5.4. *The product of natural numbers is not interpretable by their sum.* ([PRE, 1929]; stated in Volume 1, 7.1.2).

▷ Supposing the contrary, let $\mathscr{P}$ be a logical operator interpreting the product in terms of the sum. Then, for any relation S logically equivalent to the sum, the relation $\mathscr{P}(S)$ is logically equivalent to the product. Let S be the sum over the set of pairs (a, b), where a is a nonnegative rational number and b an integer (a natural number if $a=0$, an arbitrary integer if $a>0$). The sum of two pairs (a, b), (a', b') is defined as $(a+a', b+b')$ (with the usual sums of rational numbers and integers). The relation S is a positive sum in the sense of 3.4, with neutral element $(0, 0)$; it is discrete, for $(0, 1)$ is the immediate successor of $(0, 0)$. It is divisible in the sense of 3.5: given (a, b), either $a=0$, in which case everything is reduced to the case of natural numbers, or $a>0$, and then for every positive integer p we take an integer c such that $b=\underbrace{c+\cdots+c}_{p \text{ times}}+h$, where h is one of the numbers $0, 1, \ldots, p-1$, and finally consider the pair $(a/p, c)$. By 3.5.2, S is logically equivalent to the sum of natural numbers. Consider the chain $\leqslant$ associated with S (see 3.4), and let $\cdot$ denote the product $\mathscr{P}(S)$. For each positive integer p, we have

$$(1, 0)\cdot(0, p)=\underbrace{(1, 0)+\cdots+(1, 0)}_{p \text{ times}}=(p, 0)$$

Since $(1, 0)>(0, p)$ for any p, it follows that

$$(1, 0)\cdot(1, 0)>(1, 0)\cdot(0, p)=(p, 0)$$

for any p. This is a contradiction, for no element of the base can follow all pairs of the type $(p, 0)$.

3.6. Real field; elimination of quantifiers over the sum and product of algebraic numbers or real numbers

A birelation (S, P) is said to be a *(commutative) field* if S (the *sum*) and P (the *product*) are ternary relations satisfying the following conditions.

S is a commutative group; the operation is denoted by $+$, the neutral element by 0, and the inverse of a by $-a$.

P is associative, commutative, and distributive on S; the operation is denoted by $\cdot$; by the above conditions, $a\cdot 0=0$ for all a. There is a neutral element 1 for P, distinct from 0. Any element $a\neq 0$ has an inverse, denoted by $1/a$. Note that S is necessarily dense in the sense of 3.4, since for any a and any integer p there exists an element b such that $b\cdot p=a$.

A field is said to be *orderable* if -1 is not a sum of squares. It is easy to see that then no sum of squares can vanish unless all its terms vanish. We define the order *associated* with the field to be the relation $\leqslant$ such that $a\leqslant b$ if $b=a+c$, where c is a sum of squares. It is easy to see that this relation is indeed reflexive, transitive and antisymmetric. Moreover, it is compatible with the sum: if $a\leqslant b$, then $a+c\leqslant b+c$ for all a, b, c. Finally, if $a\geqslant 0$ and $b\geqslant 0$, then $a\cdot b\geqslant 0$. The field is said to be *totally orderable* if the order just defined is a chain (total ordering).

An orderable field is said to be *real* if (1) for every a there exists b such that $b^2=a$ or $b^2=-a$ (implying that the field is totally orderable), and (2) any polynomial f defined through the sum S and product P has the following property: for any a and b $(a<b)$, if $f(a)\geqslant 0$ and $f(b)\leqslant 0$, there exists an intermediate element c $(a\leqslant c\leqslant b)$ such that $f(c)=0$. A corollary of this is that a polynomial cannot pass from one value $f(a)$ to another $f(b)$ without assuming each intermediate value between a and b.

3.6.1. *Every real field is an extension of the field of algebraic numbers (up to isomorphism).*

▷ Let (S, P) be a real field with base E. We identify the integers 0 and 1 with the neutral elements of S and P, respectively. We then identify each positive rational number p/q with the unique element $a\in$ E such that

$$\underbrace{a+\cdots+a}_{q \text{ times}}=\underbrace{1+\cdots+1}_{p \text{ times}},$$

where 1 is the neutral element of P; this extends to the negative rationals too. Let u be an algebraic number; u may always be defined as the n-th root (in the natural order of the real numbers) of a polynomial f with integer coefficients, all of whose roots are simple. Let a be a rational number lying between the $(n-1)$-th and n-th roots of f, and b a rational number between the n-th and $(n+1)$-th roots. Then f may be identified with a polynomial with coefficients in E, defined in terms of S and P; a and b may be identified with two elements of E. Thus u is identified with the element of E lying between a and b such that $f(u)=0$, where 0 is the neutral element of S. ◁

3.6.2. Elimination theorem of Sturm-Tarski. Let (S, P) and (S′, P′) be two real fields with sums S, S′, products P, P′ and bases E, E′. Let a_i and a_i' $(i=1,\ldots,n)$ be two sequences of n elements of E and E′, and u, r two natural numbers. Consider a finite set of polynomials f in one variable, whose coefficients are polynomials, of the form $\sum u_j a_1^{r_{j,1}} \ldots a_n^{r_{j,n}} x^s$, where the u_j are integers of absolute value $\leqslant u$ and the exponents $r_{j,1}, \ldots, r_{j,n}$ are natural numbers such that $r_{j,1}+\cdots+r_{j,n}+s \leqslant r$. For each such polynomial f, we let f' denote the *transform* of f, i.e., the polynomial obtained from f when each a_i is replaced by a_i'. We stipulate that no given monomial $u_j a_1^{r_{j,1}} \ldots a_n^{r_{j,n}} x^s$ can occur more than once in a polynomial f (without violating the conditions $|u_j| \leqslant u$). We shall view two monomials as distinct if they are formally so; e.g., the monomials $a_1 a_2^2 x$ and $a_1^2 a_2 x$ are distinct even if $a_1=a_2$.

There exist two natural numbers u^*, r^*, depending only on u, r but not on n or on the specific values of a_i, a_i', which satisfy the following conditions:

Suppose that the two sequences a_i and a_i' $(i=1,\ldots,n)$ satisfy the same polynomial equalities and inequalities with integer coefficients of absolute value at most u^* and exponent sums at most r^*. Then:

(1) *The total number of roots x of the polynomials f is equal to the total number of roots x' of the transforms f'.*

(2) *The bijective mapping taking the roots x onto the roots x', preserving their order, maps each root of a polynomial f onto a root of the corresponding polynomial f', of the same multiplicity.*

(3) *Given any polynomial f and its transform f', two consecutive roots x and the corresponding roots x' (under the above-mentioned bijective mapping), f' has the same sign between its roots x' as f between the roots x.*

A simple example will clarify the situation. Let the set of polynomials f contain a single polynomial ax^2+bx+c (where a, b, c – the elements a_i – are real numbers); the transform is $f'=a'x'^2+b'x'+c'$ (where a', b', c' are again real numbers; S′, P′ are identical to S, P – the usual sum and product of real numbers). Take $u^*=4$, $r^*=2$. By hypothesis, b^2-4ac and $b'^2-4a'c'$ have the same sign, positive say. The total number of roots x is 2, as is the total number of roots x'; thus condition (1) is satisfied, as is condition (2) since the roots are all simple. Finally, a and a' are of the same sign, so that the sign of f between its two roots x is the same as that of f' between its roots x'; thus condition (3) also holds.

▷ (1) It will suffice to establish (1) and (2) for an arbitrary pair of polynomials f and the corresponding pair of transforms f'.

Let F be a polynomial possessing only simple roots, which are precisely the roots of the two given polynomials f. We construct a Sturm sequence $F_0, F_1, \ldots$ as follows. $F_0=F$; F_1 is the derivative of F; F_2 is minus the remainder upon division of F_0 by F_1, i.e., $F_2=-F_0+F_1G_1$ and F_2 is of degree strictly less than that of F_1; similarly, $F_3=-F_1+F_2G_2$, with the degree of F_3 strictly less than that of F_2; the procedure continues until a constant polynomial is obtained. It is readily seen that the number of changes of sign in the Sturm sequence can change only upon passage through a root of F: it is then diminished by 1 with increasing x. Thus the number of roots of F is the difference between the number of changes of sign in the Sturm sequence when the variable tends to negative infinity and the number of changes when it tends to positive infinity. These numbers, in turn, depend only on the signs of the leading coefficients of $F_0, F_1, \ldots$ and on the parity of their degrees.

These leading coefficients are polynomials in the variables a_i ($i=1, \ldots, n$) with integer coefficients; since the number of these polynomials is finite, we can speak of the maximum absolute value u^* of their coefficients, and of the maximum degree r^* of the terms of these polynomials. Note that we are concerned here with the total degree, i.e., the sum of exponents of $a_1, a_2, \ldots, a_n$. By hypothesis, when we replace each a_i by a'_i, hence replacing each f by its transform f' and the sequence $F=F_0, F_1, \ldots$ by a corresponding sequence $F' = F'_0, F'_1, \ldots$, the sign (minus, plus *or zero*) of each coefficient is preserved. The number of changes of sign when the independent variable tends to positive infinity is the same for both Sturm sequences. Thus, the number of roots of F is the same as the number of roots of F′, so that the same holds for the pairs of polynomials f and f', proving (1).

(2) The position of a root x of F relative to the other roots is entirely determined by the sequence of signs of $F_1(x), F_2(x), \ldots$, where $F_1, F_2, \ldots$ is the Sturm sequence. Now, a root x of F, of given relative position (in the natural increasing order of roots), is a root of the polynomial f if and only if the sequence of equalities and inequalities $F(x)=0$, $F_1(x) \gtrless 0, \ldots$ characterizing the root implies that $f(x)=0$. In other words, for no x can either inequality $f(x)<0$ or $f(x)>0$ be added to this sequence without entailing a contradiction. A root x of F, of given relative position, is a simple root of f if and only if the sequence of inequalities just mentioned implies that $f(x)=0$ and $f^1(x) \neq 0$ (where f^1 denotes the derivative of f). In other words, for no x can this sequence of equalities and inequalities be consistently augmented by adding either of the inequalities $f(x)<0$ or $f(x)>0$, or the pair of equalities $f(x)=0$ and $f^1(x)=0$. Similar conditions may be formulated for x to be a double, triple, etc. root of f: the condition always reduces to the statement that some finite set of polynomial equalities and inequalities, including at least the equality $F(x)=0$, is contradictory. We shall see in part 4 below how to carry these "contradictory sets" over from the polynomials f and F to their transforms f' and F′, thus proving condition (2).

(3) Any root of one of the polynomials f is a root of F; thus, in order to ascertain the sign of a polynomial f between two consecutive roots x and y of F, it will suffice to consider this sign for real numbers approaching x from above. Now a given root x of F is characterized by a sequence of equalities and inequalities $F(x)=0$, $F_1(x) \gtrless 0, \ldots$. On the other hand, the sign of f for real numbers approaching x from above is determined by the sequence of signs of $f(x), f^1(x), f^2(x), \ldots$, where $f^1, f^2, \ldots$ are the successive derivatives of f. As in part 2 above, it follows that the statement that f has a given sign between two consecutive roots of F is equivalent to a finite number of contradictions, each referring to a finite set of polynomial equalities and inequalities including (at least) $F(x)=0$.

(4) Any two polynomial equalities $A(x)=0$, $B(x)=0$ may be replaced by a single equality, for example, $(A(x))^2+(B(x))^2=0$. Thus any finite set

of polynomial equalities and inequalities including at least one equality reduces to a single equality and one or more inequalities. We shall now use the method of Lemma 2.3 in [TAR, 1940, 1967].

Given two polynomials A and B in x, we express the statement that $A(x)=0$ and $B(x)>0$ cannot both hold as follows. Using a Sturm sequence, we determine the number of roots of A which are not roots of B – this is simply the difference between the number of roots of AB and the number of roots of B. Let a^+ denote the number of roots of A that make B positive and a^- the number of roots that make B negative; we have thus determined $a^+ + a^-$. On the other hand, consider the Sturm sequence $U_0 = A$, $U_1 = BA^1$ (where A^1 is the derivative of A), $U_2 = -U_0 + U_1V_1$ (where the degree of U_2 is strictly less than that of U_1), and so on, the sequence have no common root and A has no multiple root, so that A and BA^1 have ending with a constant polynomial. We may always assume that A and B no common root. Then the number of changes of sign in this Sturm sequence when the variable tends to negative infinity, minus the number of changes when it tends to positive infinity, is precisely $a^+ - a^-$. We may thus determine a^+ and a^-, and the statement in question is equivalent to $a^+ = 0$.

Given three polynomials A, B, C in x, the statement that the sequence $A(x)=0$, $B(x)>0$, $C(x)>0$ is contradictory may be expressed as follows; the procedure extends immediately to more than two inequalities. By the method just described, we know how to determine the number of roots of A satisfying one polynomial inequality. Let a, b, c, d be respectively the numbers of roots of A satisfying the inequalities $B^2C^2>0$, $BC^2>0$, $B^2C>0$, $BC>0$. It is immediate that the number of roots of A satisfying $B>0$ and $C>0$ is $\frac{1}{2}(-a+b+c+d)$. Thus the statement that the above sequence is contradictory is equivalent to $a=b+c+d$. The method carries over easily to the case of one equality and 3, 4,... polynomial inequalities: the problem is reduced from $h \geqslant 2$ to $h-1$ inequalities.

(5) Summarizing, we have now established that each of the conditions figuring in (2) and (3) (a given root of F is or is not a root of f of given multiplicity; f is positive or negative between two given roots of F) is equivalent to a statement concerning the signs or vanishing of certain coefficients, which are themselves polynomials in the variables $a_i (i=1,\ldots, n)$ with integer coefficients. There are finitely many such polynomials for each of the conditions in question; let u^* denote the maximum absolute

value of their coefficients and r^* the maximum degree of the terms in $a_1, \ldots, a_n$. By hypothesis, the sign or vanishing of these polynomial coefficients is preserved when each a_i is replaced by a'_i. Thus all statements concerning the numbers of roots satisfying various polynomial equalities and inequalities are preserved. This proves (2) and (3) and completes the proof of the theorem. ◁

3.6.3. Let S and P denote the sum and product over the set E of algebraic numbers, and S′, P′ the sum and product of some other real field with base E′. Let n be a natural number, a_i $(i=1,\ldots,n)$ a sequence of elements of E and a'_i a sequence of elements of E′; let p be a natural number.

A sufficient condition for the mapping taking each a_i onto a'_i $(i=1,\ldots,n)$ to be a (p, p)-isomorphism of (S, P) *onto* (S′, P′) *is that the elements a_i and a'_i satisfy the same equalities and inequalities* (mod S *or* mod S′) *of the form*

$$\sum u_j a_1^{r_{j,1}} \ldots a_n^{r_{j,n}} = 0, \qquad >0, \qquad <0,$$

where the coefficients u_j are integers of absolute value $\leqslant u(p)$, the exponents r_j are natural numbers such that $r_{j,1}+\cdots+r_{j,n}\leqslant r(p)$.

The functions u and r are defined by $u(0)=r(0)=2$ and, for each p,

$$u(p+1)=u^* \quad \text{and} \quad r(p+1)=r^*,$$

where u^* and r^* are the numbers whose existence is proved in the elimination theorem, for the (finite) set of all polynomials in one variable whose coefficients are polynomials in the a_i $(i=1,\ldots,n)$ or a'_i with integer coefficients of absolute value $\leqslant u(p)$ and exponents such that $r_1+\cdots+r_n+ +s\leqslant r(p)$ (s is the exponent of the variable).

▷ The assertion is true for $p=0$, $u(0)=r(0)=2$. Indeed, the mapping taking each a_i onto a'_i $(i=1,\ldots,n)$ will be a local isomorphism of (S, P) onto (S′, P′) if all equalities and inequalities of the following types are preserved upon passage from (S, P) and the a_i to (S′, P′) and the a'_i: $a_i=0$, $a_i+a_j-a_k=0$, $2a_i-a_j=0$, $a_i-1=0$ (this condition arises from $a_i^2=a_i$, $a_i\neq 0$), $a_ia_j-a_k=0$, $a_i^2-a_j=0$; in addition, all inequalities obtained when $=$ is replaced by $>$ and $<$ $(i, j, k=1,\ldots,n)$.

Supposing the assertion true for p, we now prove it for $p+1$. The mapping taking each a_i onto a'_i $(i=1,\ldots,n)$ will be a $(p+1, p+1)$-isomorphism of (S, P) onto (S′, P′) if for every $b\in E$ there exists $b'\in E'$ such that the mapping $a_i\to a'_i$, $b\to b'$ is a (p, p)-isomorphism, and conversely with a_i and

a'_i interchanged. It will thus suffice to show that for every b there exists b' such that $a_1, \ldots, a_n$, b, on the one hand, and $a'_1, \ldots, a'_n$, b', on the other, satisfy the same equalities and inequalities of the form

$$\sum u a_1^{r_1} \cdots a_n^{r_n} b^s = 0, \qquad > 0, \qquad < 0,$$

where the u's are integers of absolute value $\leqslant u(p)$, the r's and s's range over all integers such that $r_1 + \cdots + r_n + s \leqslant r(p)$ (and conversely). Consider the finite set of polynomials f on the left of these equalities and inequalities, with b viewed as their only variable. To these polynomials we apply the elimination theorem, with $u(p+1) = u^*$ and $\mathrm{r}(p+1) = r^*$. Replacing each a_i by a'_i $(i = 1, \ldots, n)$, we obtain polynomials f' whose total number of roots is equal to the total number of roots of the polynomials f. The bijective mapping of these roots b onto the roots b' of the polynomials f', which preserves their order, maps the roots of each f onto the roots of its transform f' in such a way that f and f' have the same sign between pairs of corresponding consecutive roots. It therefore suffices to consider the interval containing the given element b and to take an element b' in the corresponding interval; the result will be polynomials f' satisfying the same equalities and inequalities as the polynomials f. The same argument holds with (S, P) and (S′, P′), a_i and a'_i interchanged. ◁

3.6.4. *Any two real fields* (S, P) *and* (S′, P′) *are logically equivalent. Moreover, if* (S′, P′) *is an extension of* (S, P), *then it is a logical extension of* (S, P).

▷ Suppose first that (S′, P′) is an extension of (S, P). Given any finite sequence of elements a_i of the base of (S, P), the conditions of 3.6.3 are satisfied with $a'_i = a_i$ $(i = 1, \ldots, n)$. Thus the identity mapping on any finite subset of the base is a (p, p)-isomorphism of (S, P) onto (S′, P′) for every p; it follows that (S′, P′) is a logical extension of (S, P).

Now we saw in 3.6.1 that every real field is an extension (up to isomorphism) of the field of algebraic numbers; thus any two real fields are logically equivalent. ◁

3.6.5. *Neither the set of natural numbers nor the set of integers is interpretable by the sum and product of algebraic numbers, of real numbers, or by any real field* ([TAR, 1940, 1951]; see Volume 1, 7.3.5).

▷ Suppose that the unary relation E which is true, say, for the integers alone is interpretable in (S, P), where S and P are the sum and product of real numbers or, more generally, the sum and product of some real field. By Volume 1, 7.3.5, the unary relation which is true for the algebraic numbers is interpretable in (S, P, E), and so, by hypothesis, in (S, P). The corresponding formula A is such that $\exists \neg A(x)\,(S, P) = +$, unless the base consists of the algebraic numbers alone, in which case the value is $-$; this contradicts the fact that any real field is logically equivalent to the field of algebraic numbers (see 3.6.4). $\triangleleft$

EXERCISES

1

Let S be a commutative group. S is totally orderable if and only if it is torsionfree: $a+\cdots+a=0$ (mod S) implies $a=0$, for any a. Necessity is obvious; prove sufficiency. *Hint:* Let A be any partial order over the same base as S, obtained by partitioning the base into classes and stipulating that two elements are comparable (mod A) if and only if they belong to the same class. Assume that A is compatible with S: if $x \leqslant y$, then $x+z \leqslant y+z$ (mod A) for any x, y, z. Show that any partial order A which is not a total order may be extended to another partial order.

2

(1) Following the lines of 3.4.3, show that the sum of all rationals is not finitely-axiomatizable. Do the same for the sum of all integers.

(2) Extend the results of 3.4.2 to the birelation "sum, comparison" of rationals of arbitrary sign; extend 3.5.2 to the birelation "sum, comparison" of integers. Hence deduce that these birelations are not finitely-axiomatizable.

Problem. Is the unary relation which is true for the positive integers alone interpretable in the birelation "sum of integers, singleton 1"?

3

By Theorem 3.6.4, the field of real numbers is a logical extension of the field of algebraic numbers. Show that if a finite system of equations and inequalities with rational coefficients is solvable in real numbers, it is solvable in algebraic numbers.

4

There exist totally orderable sums which are neither groups nor positive sums. For example, consider the set of ordered pairs (a, b), where a is a natural number and b any integer; define $(a, b)+(a', b')$ to be $((a+a'), (b+b'))$. Show that this is a totally orderable sum in the sense of 3.4. Note that the set of pairs (a, b), where both a and b are natural numbers, is not totally orderable (take $a' > a$, $b' < b$).

Problems. (1) Given any totally orderable sum, can we express any element as $a+b$, where a is in some positive sum and b in some commutative group?

(2) Given any torsionfree sum (see Exercise 1), does there exist a total ordering which is compatible with it?

CHAPTER 4

EXTENSION THEOREMS

4.1. Restrictive sequence; (k, p)-isomorphism and (k, p)-identimorphism

Let R be a multirelation with base E. Consider an infinite decreasing sequence of subsets of E indexed by the natural numbers, $E_0 \supseteq E_1 \supseteq E_2 \supseteq \supseteq \cdots$. The subsets E_i $(i=0, 1, 2, \ldots)$ may be empty from some i onward. The sequence of restrictions $R_i = R \mid E_i$ will be called a *restrictive sequence* for R. We shall identify R itself with the restrictive sequence defined by $E_i = E$ $(i=0, 1, 2, \ldots)$.

4.1.1. Let R and R′ be multirelations with bases E and E′, respectively. Consider a restrictive sequence R_i with bases $E_i \subseteq E$, and similarly a restrictive sequence R'_i with bases $E'_i \subseteq E'$. Let f be a bijective mapping of a subset $F \subseteq E$ onto $F' \subseteq E'$. If $F \subseteq E_0$ and $F' \subseteq E'_0$, and f is a local isomorphism of R onto R′, we shall say that for each natural number p the mapping f is a $(0, p)$-*isomorphism* of the sequence R_i onto the sequence R'_i. Let $k \geqslant 1$ and suppose we have already defined the concept of $(k-1, p)$-isomorphism for each p. Then f will be a (k, p)-isomorphism of the sequence R_i onto R'_i if $F \subseteq E_k$, $F' \subseteq E'_k$, and for all $q \leqslant p$ and any subset F^* obtained from F by adding q elements of E_{k-1}, there exists an extension of f to F^* which is a $(k-1, p-q)$-isomorphism of the sequence R_i onto the sequence R'_i; and conversely, with F, E_{k-1}, f, R_i, R'_i replaced by F′, E'_{k-1}, f^{-1}, R'_i, R_i, respectively.

A (k, p)-*automorphism* of a sequence R_i is a (k, p)-isomorphism of the sequence onto itself.

The inverse of a (k, p)-isomorphism and the product of two (k, p)-isomorphisms are also (k, p)-isomorphisms. The identity mapping on a subset of E_k is a (k, p)-automorphism of the sequence for any p. Any isomorphism which transforms each R_i into R'_i, restricted to a subset of E_k, is a (k, p)-isomorphism for any p. Any (k, p)-isomorphism restricted to a subset of its domain of definition is a (k, p)-isomorphism. Finally, any

(k, p)-isomorphism is also a (k', p')-isomorphism for any $k' \leqslant k$, $p' \leqslant p$.

A restrictive sequence is said to be (k, p)-*equivalent* to another if the empty mapping is a (k, p)-isomorphism of the first onto the second; it follows from the preceding assertions that this relation is reflexive, symmetric and transitive.

Identifying each multirelation R with the restrictive sequence all of whose terms are identical to R, we have simply (k, p)-isomorphism and (k, p)-equivalence of multirelations (see 1.1, 1.2).

4.1.2. Two restrictive sequences R_i and R'_i are said to be (k, p)-*identimorphic* if, for every $i \leqslant k$, the identity mapping on $E_i \cap E'_i$ (where E_i and E'_i are the bases of R_i and R'_i, respectively) is an (i, p)-isomorphism of the first sequence onto the second, By the preceding properties, this relation is reflexive and symmetric. Moreover, (k, p)-identimorphism implies (k, p)-equivalence. However, (k, p)-*identimorphism is not transitive.* Indeed, consider three elements a, b, c and the unary relations A, B, C with bases $\{b, c\}$, $\{c, a\}$, $\{a, b\}$ defined by

$$A(b)=A(c)=+, \quad B(a)=-, \quad B(c)=+, \quad C(a)=C(b)=-.$$

Then A and B are 0-identimorphic (i.e., $(0, p)$-identimorphic for any p), for they have the same restriction to the intersection of their bases (which is the single element c). Similarly, B and C are 0-identimorphic. However, A and C are not identimorphic, since $A(b)=+$ and $C(b)=-$.

If two restrictive sequences coincide up to their terms R_k, they are (k, p)- identimorphic for any p.

Let R *be a multirelation. Let* R_i *and* R'_i *be restrictive sequences for* R. *If each of these sequences is* (k, p)-*identimorphic to* R *(i.e., to the sequence all of whose terms are* R*), then they are* (k, p)-*identimorphic to each other.*

$\triangleright$ Let k, p be two natural numbers, E_k the base of R_k and E'_k that of R'_k. By hypothesis, the identity mapping on $E_k \cap E'_k$ is a (k, p)-isomorphism of the sequence $\{R_i\}$ onto R and also of R onto the sequence $\{R'_i\}$. $\triangleleft$

4.2. Application to Logical Restriction

4.2.1. *Let* R *be a multirelation, k and p two natural numbers, and* H *a finite subset of the base of* R. *There exists a restrictive sequence with finite bases* $F_i (i=0, 1, 2, \ldots)$, $F_k = H$, *which is* (k, p)-*identimorphic to* R. *Moreover, the cardinality of* F_0 *may be bounded by a number depending only on the arity of* R, *the numbers k, p and the cardinality of* H.

▷ For each $q \leqslant p$, we partition the sequences of q elements from $E-H$ into $(k-1, p-q)$-equivalence classes: two sequences are said to be equivalent if the identity mapping on H, extended by mapping one sequence onto the other, is a $(k-1, p-q)$-automorphism of R. The number of these classes is finite and depends only on the arity of R, on k and p, and on the cardinality of H (see 1.2.1). Take one representative sequence from each class and define F_{k-1} $(k \geqslant 1)$ as the set of values of the sequences for $q=0, 1, \ldots, p$. If $k \geqslant 2$, we iterate the procedure, replacing $F_k = H$ by F_{k-1} and considering $(k-2, p-q)$-equivalence classes; we thus construct the subsets F_i $(i=k, k-1, \ldots, 0)$ in succession. It remains to show that for each $k' \leqslant k$ the identity mapping on $F_{k'}$ is a (k', p)-isomorphism of R onto the restrictive sequence. This is obvious for $k'=0$. Let $k' \geqslant 1$ and suppose the assertion true for $k'-1$. For any sequence s of $q \leqslant p$ elements of $F_{k'-1}$, the identity mapping on the union of $F_{k'}$ and the elements of s is by assumption a $(k'-1, p)$-isomorphism, hence also a $(k'-1, p-q)$-isomorphism of the restrictive sequence onto R. Conversely, for any sequence s comprising $q \leqslant p$ elements of the base of R, there exists by hypothesis a sequence t of elements of $F_{k'-1}$ such that the identity mapping on $F_{k'}$, extended by the mapping taking s onto t, is a $(k'-1, p-q)$-automorphism of R. Taking the product of the latter with the identity mapping on t, viewed as a $(k'-1, p)$-isomorphism of R onto the restrictive sequence, we get a $(k'-1, p-q)$-isomorphism of R onto the restrictive sequence. ◁

4.2.2. *Let* R *be a multirelation, k and p two natural numbers,* $\mathscr{A}$ *a set of restrictive sequences of* R, *any two of which are* (*k*, *p*)-*identimorphic. Suppose that for any finite subset* H *of the base* |R| *there is a restrictive sequence in* $\mathscr{A}$ *whose k-th base contains* H. *Then every restrictive sequence in* $\mathscr{A}$ *is* (*k*, *p*)-*identimorphic to* R.

▷ The assertion is true for $k=0$, for any restrictive sequence is $(0, p)$-identimorphic to R for any p. Let $k \geqslant 1$ and suppose the assertion true for $k-1$, so that every restrictive sequence in $\mathscr{A}$ is $(k-1, p)$-identimorphic to R. Let F_i $(i=0, 1, \ldots, k)$ be the bases of a restrictive sequence S in $\mathscr{A}$.

We claim that the identity mapping on F_k is a (k, p)-isomorphism of S onto R. On the one hand, for any subset H obtained by adding $q \leqslant p$

elements of F_{k-1} to F_k, the identity mapping on H is by hypothesis a $(k-1, p)$-isomorphism, hence also a $(k-1, p-q)$-isomorphism of S onto R. On the other hand, for any subset H obtained by adding $q \leqslant p$ elements of $|R|$ to F_k, there exists by hypothesis a restrictive sequence T in $\mathscr{A}$, with bases G_i $(i=0, 1, \ldots, k)$, such that $H \subseteq G_k$. By hypothesis, S and T are (k, p)-identimorphic; thus the identity mapping on $F_k = F_k \cap G_k$ is a (k, p)-isomorphism of T onto S. Since $H \subseteq G_k \subseteq G_{k-1}$, so that H is the union of $F_k \cap G_k$ and a set of q elements of G_{k-1}, it follows that there exists a bijective mapping f of H into F_{k-1}, whose restriction to F_k is the identity mapping, which is a $(k-1, p-q)$-isomorphism of T onto S. Taking the product of f and the identity mapping on H (which is by assumption a $(k-1, p)$-isomorphism, hence also a $(k-1, p-q)$-isomorphism of R onto T), we see that f is a $(k-1, p-q)$-isomorphism of R onto S. ◁

4.2.3. *Let* R *be a multirelation,* D *a subset of* $|R|$. *Suppose that for every finite subset* H *of* D *and any natural numbers k, p there exists a restrictive sequence of* R, *with bases* F_i $(i=0, 1, \ldots, k)$ *such that* $H \subseteq F_k \subseteq \cdots \subseteq F_0 \subseteq D$, *which is* (k, p)*-identimorphic to* R. *Then* $R \mid D$ *is a logical restriction of* R.

▷ Let H be a finite subset of D; we claim that the identity mapping on H is a (k, p)-isomorphism of R onto the restriction $R \mid D$. Fixing k and p, we associate with each finite subset H of D a restrictive sequence S_H whose k-th base contains H, which is (k, p)-identimorphic to R. By 4.1.2, if we allow H to vary, the resulting sequences S_H are pairwise (k, p)-identimorphic. By 4.2.2, each of them is (k, p)-identimorphic to $R \mid D$. Thus the identity mapping on H is both a (k, p)-isomorphism of R onto S_H and of S_H onto $R \mid D$, hence of R onto $R \mid D$. ◁

4.2.4. We can now offer another proof of the logical restriction theorem (see 1.4.4):

For any infinite subset D *of the base* $|R|$, *there exists a superset* D^* *of* D, *equipollent to* D, *such that* $R \mid D^*$ *is a logical restriction of* R (assuming the axiom of choice).

▷ For every finite subset H of D and every pair k, p, we take a restrictive sequence with finite bases whose k-th base contains H, and which is (k, p)-identimorphic to R (see 4.2.1). The union of the 0-th bases is a superset D_1 of D which is equipollent to D. Iterating the construction, we similarly obtain supersets $D_2 \supseteq D_1, \ldots, D_{i+1} \supseteq D_i$, for each i. The union

D^* of the sets D_i is a superset of D which is equipollent to D and satisfies the hypotheses of 4.2.3 (with D replaced by D^*). ◁

4.3. Projection filter

Let E and E′ be two sets and f a mapping of E into E′. Let m be a natural number, R and R′ two m-ary relations with bases E and E′, respectively. R is said to be the *inverse projection of* R′ *under* f if, for any elements $a_1, \ldots, a_m$ of E,

$$R(a_1, \ldots, a_m) = R'(f(a_1), \ldots, f(a_m));$$

we write $R = f^{-1}(R')$. If f is bijective, it is an isomorphism of R onto R′ (see Volume 1, 3.2).

Fixing the sets E and E′, consider all mappings f of E into E′. Any filter defined on these mappings is called a *projection filter* of E onto E′. The relation R is called the *inverse projection of* R′ *under the filter* $\mathscr{F}$, $R = \mathscr{F}^{-1}(R')$, if, for all elements $a_1, \ldots, a_m \in E$, the set of mappings f such that

$$R(a_1, \ldots, a_m) = R'(f(a_1), \ldots, f(a_m))$$

is an element of $\mathscr{F}$, in other words, if the equality holds for *almost every mapping* of $\mathscr{F}$.

For given R′ and $\mathscr{F}$, there exists at most one inverse projection: this follows from the fact that the intersection of two elements of the filter is nonempty.

If $\mathscr{F}$ and $\mathscr{G}$ are two projection filters of E onto E′, $\mathscr{G}$ is finer than $\mathscr{F}$ (i.e., $\mathscr{G}$ contains $\mathscr{F}$), and the inverse projection under $\mathscr{F}$ exists, then the latter is also the inverse projection under $\mathscr{G}$.

4.3.1. *If $\mathscr{F}$ is an ultrafilter, the inverse projection exists (and is unique).*

For any set of mappings, either the set itself or its complement is in $\mathscr{F}$; hence either $R'(f(a_1), \ldots, f(a_m)) = +$ for almost every mapping, or $R'(f(a_1), \ldots, f(a_m)) = -$ for almost very mapping.

If the ultrafilter $\mathscr{F}$ is trivial, i.e., consists of all sets containing a single mapping f, the inverse projection under $\mathscr{F}$ coincides with the inverse projection under f.

An application of the projection filter was described in Volume 1, 3.2.6.

4.3.2. EQUIVALENT ELEMENTS, INJECTIVE FILTER. Consider a projection filter of E onto E′. Two elements a, $b \in$E are said to be *equivalent* under the filter if $f(a)=f(b)$ for almost every mapping f. The relation thus defined is reflexive, symmetric and transitive, since the intersection of two elements of a filter is again in the filter.

A projection filter of E onto E′ is said to be *injective* if, for any two distinct elements a, b of E, we have $f(a) \neq f(b)$ for almost every f. With each finite subset F of E we associate the set U_F of all mappings of E into E′ whose restrictions to F are injective. Then the family $\mathscr{F}$ of all sets U_F and all their supersets constitutes an injective filter. A projection filter is injective if and only if it is finer than $\mathscr{F}$. It can be shown that for an infinite set E′ and any set E there exists an injective filter of E onto E′. Any filter which is finer than an injective filter is injective.

If $\mathscr{F}$ is the trivial ultrafilter generated by a single mapping f, then $\mathscr{F}$ is injective if and only if f is injective; this implies that the cardinality of E cannot exceed that of E′.

An ultrafilter is injective if and only if no two distinct elements of E are equivalent.

4.3.3. *Let* R *and* R′ *be relations of the same arity, with bases* E *and* E′, *respectively. There exists an injective projection filter* $\mathscr{F}$ *such that* R= $=\mathscr{F}^{-1}$(R′) *if and only if, for every finite subset* F *of* E, *the restriction* R | F *is isomorphic to a restriction of* R′.

▷ Let m be the arity of R and R′. Suppose that R$=\mathscr{F}^{-1}$(R′) and let F be a finite subset of E. The set of mappings f whose restrictions to F are bijective is an element of $\mathscr{F}$. Divide this element into two subsets, according as $R'(f(a_1),\ldots, f(a_m))$ is + or −. One of these subsets is an element of $\mathscr{F}$, and any of its members maps R | F onto a restriction of R′ (the elements $a_1,\ldots, a_m$ run through F). Conversely, suppose that for every finite subset F of E there exists a mapping f of F into E′ which is an isomorphism of R | F onto a restriction of R′. Let U_F denote the (non-empty) set of all mappings f of E into E′ whose restrictions to f are isomorphisms of this type. Then the family of all these sets U_F and their supersets consitutes a projection filter of E onto E′ under which R is the inverse projection of R′. ◁

4.3.4. IDENTICAL FILTER. Let E and E′ be two sets and D a subset of the intersection $E \cap E'$. A projection filter of E onto E′ is said to be D-*identical* if it contains an element consisting solely of mappings of E into E′ whose restriction to D is the identity mapping. In other words, almost every mapping is the identity mapping on D.

Given relations R and R′ with bases E and E′, respectively, if $\mathscr{F}$ is a D-identical filter and $R = \mathscr{F}^{-1}(R')$, then the restrictions $R \mid D$ and $R' \mid D$ coincide.

If E′ is infinite and $E \supseteq E'$, there exists a filter of E onto E′ which is E′-identical and injective. Indeed, for each finite subset F of E, let U_F be the set of all bijective mappings of F into E′ whose restriction to $F \cap E'$ is the identity mapping. Then the family of all supersets of the sets U_F is the required filter.

4.3.5. Application to 1-extensions. Let R be a relation with base E and E* a superset of E. A relation R* with base E* is said to be a 1-*extension* of R if, for every finite subset F of E, the identity mapping on F is a $(1, p)$-isomorphism of R onto R* for every p. An equivalent definition is as follows: for any finite subset G of E*, there exists an isomorphism of $R^* \mid G$ onto a restriction of R which reduces to the identity mapping on $G \cap E$ (see Volume 1, Chapter 3, Exercise 7). For example, the chain obtained by adding one element "at the end" of the chain of integers is a 1-extension of the latter.

Any 1-extension of R has the same finite restrictions as R (up to isomorphism). The converse is false: if we add one element "before" 0 in the chain of nonnegative integers, we obtain an extension which has the same finite restrictions (up to isomorphism) but is not a 1-extension.

If R* is a 1-extension of R and R** a 1-extension of R*, then R** is a 1-extension of R.

Let R *be a relation with base* E *and* R* *a relation of the same arity whose base* E* *is a superset of* E. *Then* R* *is a* 1-*extension of* R *if and only if there exists a projection filter* $\mathscr{F}$ *of* E* *onto* E *which is injective,* E-*identical, and such that* $R^* = \mathscr{F}^{-1}(R)$.

This yields an alternative proof of 4.3.3, with the additional conclusion that the mappings f whose restrictions to F are bijective reduce to the identity mapping on $E \cap F$.

4.3.6. Restriction and extension of a filter. Let $\mathscr{F}$ be a projection filter of E onto E′ and D a subset of E. With each element U of $\mathscr{F}$ (which is a set of mappings f of E into E′) we associate the set U | D of restrictions of the mappings f to D. The family of all sets U | D constitutes a projection filter of D onto E′, known as the *restriction* of $\mathscr{F}$ to D and denoted by $\mathscr{F} \mid D$. If R and R′ are relations with bases E and E′, respectively, and $R = \mathscr{F}^{-1}(R')$, then $R \mid D = (\mathscr{F} \mid D)^{-1}(R')$.

If D′ is a subset of D, then $\mathscr{F} \mid D' = (\mathscr{F} \mid D) \mid D'$.

If the filter $\mathscr{F}$ is injective, then any restriction of $\mathscr{F}$ is injective. If $\mathscr{F}$ is an ultrafilter, then any restriction of $\mathscr{F}$ is an ultrafilter.

Let E be a set, E* a superset of E. Given a set U of mappings of E into E′, we define the *canonical extension* of U to E* as the set of all mappings of E* into E′ whose restrictions to E are in U. Given a filter $\mathscr{F}$, the set of canonical extensions of the members U of $\mathscr{F}$ and their supersets constitutes a projection filter of E* onto E′, known as the *canonical extension* of $\mathscr{F}$ to E*.

Let $\mathscr{F}$ be a projection filter of E onto E′ and E* a superset of E; let $\mathscr{F}^*$ be a projection filter of E* onto E′. Then we call $\mathscr{F}^*$ an *extension* of $\mathscr{F}$ to E* if $\mathscr{F} = \mathscr{F}^* \mid E$.

$\mathscr{F}^*$ is an extension of an ultrafilter $\mathscr{F}$ to E* if and only if $\mathscr{F}^*$ is finer than the canonical extension of $\mathscr{F}$ to E*.

If $\mathscr{F}^*$ is an extension of $\mathscr{F}$ and $\mathscr{F}^{**}$ an extension of $\mathscr{F}^*$, then $\mathscr{F}^{**}$ is an extension of $\mathscr{F}$. If $\mathscr{F}$ is a projection filter of E onto E′, E* a superset of E, E** a superset of E*, and $\mathscr{G}$ an extension of $\mathscr{F}$ to E**, then the restriction $\mathscr{G} \mid E^*$ is an extension of $\mathscr{F}$ to E*. Let $\mathscr{F}$ be a projection ultrafilter of E onto E′, E* a superset of E, and $\mathscr{F}^*$ an extension of $\mathscr{F}$ to E*; then any ultrafilter which is finer than $\mathscr{F}^*$ is an extension of $\mathscr{F}$ to E*.

4.3.7. LIMIT OF A SEQUENCE OF FILTERS. Suppose that for each $i = 0, 1, \ldots$ we have a set E_i, $E_i \subseteq E_{i+1}$, and let E′ be a set. For each i, let $\mathscr{F}_i$ be a projection filter of E_i onto E′, such that $\mathscr{F}_{i+1}$ is an extension of $\mathscr{F}_i$. Let D be the union of the sets E_i; with each i and each element U of $\mathscr{F}_i$ we associate the canonical extension U* of U to D (see 4.3.6). Then the family of all supersets of the sets U* constitutes a projection filter of D onto E′, known as the *limit* of the filters $\mathscr{F}_i$.

If all the $\mathscr{F}_i$ are injective filters, their limit is also injective.

The limit of a sequence of filters $\mathscr{F}_i$ is an extension of each $\mathscr{F}_i$.

▷ For each element U of a filter $\mathscr{F}_i$, the canonical extension of U to D is an element of the limit. Conversely, for any integer i and element V of the limit, there exist an integer j and an element W of $\mathscr{F}_j$ such V is a superset of the canonical extension of W to D. We distinguish two cases: (a) If $j \geqslant i$, then $\mathscr{F}_j$ is an extension of $\mathscr{F}_i$, so that the restriction of W to E_i is an element of $\mathscr{F}_i$; thus the same is true of the restriction of V to E_i. If $j < i$, then $\mathscr{F}_i$ is an extension of $\mathscr{F}_j$; the canonical extension W_i of W to E_i is an element of $\mathscr{F}_i$, whose canonical extension to D is precisely V. ◁

Any projection filter of D onto E′ is a common extension of all the filters $\mathscr{F}_i$ if and only if it is finer than the limit of the $\mathscr{F}_i$.

4.4. Logical extension theorems

4.4.1. Let E and E′ be two infinite sets, $\mathscr{F}$ an injective projection ultrafilter of E onto E′, and R′ a multirelation with base E′.

Let E^* *be a superset of* E *such that the cardinal of* $E^* - E$ *is at least that of* E. *There exists an injective projection ultrafilter* $\mathscr{F}^*$ *extending* $\mathscr{F}$ *to* E^* *and satisfying the following condition:*

For any finite subset H *of* E^* *and any k and p, there exists a sequence of finite subsets* H_i *of* E^*, *with* $H_k \subseteq H$, *whose image under almost any mapping* (mod $\mathscr{F}^*$) *is a restrictive sequence which is* (k, p)*-identimorphic to* R′.

▷ We first prove the assertion for the special case that H is a finite subset of E, instead of E^*; H, k and p will be fixed until further notice. For each mapping f of E into E′, consider the image $f(H)$. With each f we associate a restrictive sequence of R′ whose k-th base is $f(H)$ and which is (k, p)-identimorphic to R′ (see 4.2.1); let $F'_i(f)$ $(i=0, 1, \ldots, k)$ be the first $k+1$ bases of this sequence. Since the cardinality of $f(H)$ is at most that of H, it follows from 4.2.1 that the cardinality of $F'_0(f)$ is bounded by a fixed number, independent of f. Since $\mathscr{F}$ is an ultrafilter and the sequence of cardinalities of the sets $F'_i(f)$ $(i=0, 1, \ldots, k)$ may take only values bounded by a fixed integer, there exists a unique sequence of natural numbers a_i $(i=0, 1, \ldots, k)$ such that for almost every f (mod $\mathscr{F}$) the cardinality of $F'_i(f)$ is a_i. Since $\mathscr{F}$ is injective, almost every f is injective on H, so that $a_k = \text{card}(H)$.

With each $i=0, 1, \ldots, k$, we now associate the set G_i of all elements $u \in E$ such that $f(u) \in F'_i(f)$ for almost every f. For each $i=0, 1, \ldots, k$, the cardinality of G_i is a natural number $a_i' \leqslant a_i$, since $\mathscr{F}$ is injective and so almost every mapping f is injective on G_0. Let D be a finite subset of $E^* - E$ with $a_0 - a_0'$ elements and set $F_0 = G_0 \cup D$; then F_0 is of cardinality a_0. Similarly, taking $a_1 - a_1'$ arbitrary elements of $F_0 - G_0$ and considering their union with G_1, we get a set F_1 of cardinality a_1; this procedure continues till we obtain the set $F_k = G_k = H$ of cardinality $a_k = a_k'$. With each mapping f of E into E′ we associate an extension f^*, mapping $E \cup F_0$ into E′, satisfying the condition: if f maps G_i injectively into $F_i'(f)$ (this is true for almost every f), then the associated mapping f^* is injective on F_0 and maps F_i onto $F_i'(f)$ $(i = 0, 1, \ldots, k)$.

We now vary H, k and p; in accordance with the above construction, each triple (H, k, p) is associated with a sequence of sets $F_i(H, k, p)$ and $G_i(H, k, p)$, $i = 0, 1, \ldots, k$. We take the precaution of choosing the differences $F_0(H, k, p) - G_0(H, k, p)$ in such a way that they are pairwise disjoint for distinct triples (H, k, p). This implies that the union E^* of the sets F_0 and the union $E^* - E$ of the sets $F_0 - G_0$ are at most equipollent to E. Moreover, we may thus extend each f to a mapping f^* of E^* into E′ such that, whenever f maps each set $G_i(H, k, p)$ into $F_i'(f)$ (and for any given triple (H, k, p) this is true for almost every f), the extension f^* is injective on $F_0(H, k, p)$ and maps each set $F_i(H, k, p)$ onto $F_i'(f)(H, k, p)$, $i = 0, 1, \ldots, k$. If f is not injective on a given set $G_0(H, k, p)$, then f^* is arbitrary on the corresponding set $F_0(H, k, p)$.

With each element U of the ultrafilter $\mathscr{F}$ we associate the set U^* of mappings f^* associated with the members of U. The family of all supersets of the sets U^* constitutes a projection ultrafilter $\mathscr{F}^*$ of E^* onto E′ which is an extension of $\mathscr{F}$ to E^*. The filter $\mathscr{F}^*$ need not be injective, but two elements of E^* which are equivalent mod $\mathscr{F}^*$ cannot both belong to E, neither can they belong to $F_0(H, k, p)$ for the same triple (H, k, p). Take one representative of each of these equivalence classes in $E^* - E$, retaining the notation E^* for the smaller set thus obtained, which is still a superset of E, and the notation $\mathscr{F}^*$ for the restriction of the ultrafilter to E^*, which is now an *injective* projection ultrafilter of the new E^* onto E′, still an extension of $\mathscr{F}$. Now, for any finite subset H of E and any k, p, the sequence of bases $F_i(H, k, p)$ $(i = 0, 1, \ldots, k)$ is mapped by almost every mapping f^* (mod $\mathscr{F}^*$) associated with a mapping f onto the restrictive sequence of R′ with bases $F_i'(f)$ (H, k, p), which is (k, p)-identimorphic to R′.

To complete the proof for our special case (finite subsets H of E), it remains to verify that the ultrafilter $\mathscr{F}^*$ always exists, provided that E^*-E is at least equipollent to E. In the preceding argument, we obtained $\mathscr{F}^*$ as an extension of $\mathscr{F}$ to a set E^* such that the cardinality of E^*-E is at most that of E; if $\text{card}(E^*-E)<\text{card}(E)$, this may only be because certain equivalent elements were dropped in order to render $\mathscr{F}^*$ injective. If the situation *a priori* is that E^*-E is at least equipollent to E, we construct an ultrafilter $\mathscr{F}_1^*$ as before, using instead of E^* a subset E_1^* of E^* which contains E; we then let $\mathscr{F}^*$ be an injective ultrafilter extending $\mathscr{F}_1^*$ to E^*: this is always possible, for we need only take the canonical extension of $\mathscr{F}^*$ (see 4.3.6) and, for each finite subset F of E^*, add the set of mappings of E^* into E' which are injective on F. The condition stipulated in our assertion remains valid when we replace $\mathscr{F}_1^*$ by its extension $\mathscr{F}^*$.

We now proceed to the general case. We first set $E_0=E$ and construct a superset E_1 of E_0, equipollent to E^*, and an injective ultrafilter $\mathscr{F}_1$, extending $\mathscr{F}$ to E_1, which satisfies the required condition for finite subsets H of E_0. Repeating this procedure, we ultimately obtain an infinite sequence of supersets E_i $(i=0, 1,\ldots)$ and, for each of them, a projection ultrafilter $\mathscr{F}_i$ of E_i onto E' such that $\mathscr{F}_{i+1}$ is an extension of $\mathscr{F}_i$ for each i. Let E^* be the union of the sets E_i and let $\mathscr{F}$ be the corresponding limit filter, which is an injective filter of E^* onto E' (see 4.3.7). Refining $\mathscr{F}$ (if necessary), we obtain an ultrafilter $\mathscr{F}^*$ which is an extension of each $\mathscr{F}_i$.

For each finite subset H of E^*, there exists i such that $H\subseteq E_i$; thus the required condition is satisfied if we first replace $\mathscr{F}^*$ by $\mathscr{F}_{i+1}$, and it will remain valid upon returning to $\mathscr{F}^*$, since the latter is an extension of all the $\mathscr{F}_i$.

It remains to observe that, in view of the special case considered above, we may assume that E_1 and E_1-E are from the start equipollent to E^*, and subsequently that each of the differences $E_{i+1}-E_i$ $(i=1, 2,\ldots)$ is equipollent to E^*. $\lhd$

4.4.2. Let R be a multirelation with base E and R′ a multirelation of the same arity with base E′. Suppose there exists an injective projection filter $\mathscr{F}$ such that $R=\mathscr{F}^{-1}(R')$.

(1) *Let k, p be two natural numbers. Assume that, for any finite subset* H *of* E, *there exists a restrictive sequence of* R *with finite bases whose k-th*

base contains H, *with the property: the image of the sequence under almost every mapping* (mod$\mathscr{F}$) *is* (k, p)*-identimorphic to* R′. *Then for any* H *the restriction of almost every mapping to* H *is a* (k, p)*-isomorphism of* R *onto* R′.

(2) *If the above assumption is valid for all* k *and* p, *then* R *and* R′ *are logically equivalent.*

(3) *If, moreover,* E′ *is a subset of* E *and the filter* $\mathscr{F}$ *is* E′*-identical, then* R′ *is a logical restriction of* R.

▷ (1) Let k, p be fixed natural numbers. Let $\mathscr{A}$ be the set of restrictive sequences obtained by letting H range over all finite subsets H of E. Any two restrictive sequences in $\mathscr{A}$, with bases F_i and G_i $(i=0, 1,\ldots)$, are transformed by almost every mapping f into two restrictive sequences with bases $f(F_i)$ and $f(G_i)$. Each of the latter sequences is (k, p)-identimorphic to R′, so that each is (k, p)-identimorphic to the other. Moreover, since the filter is injective, almost every mapping f is injective on $F_0 \cup G_0$, which is a finite set. Hence the restriction of almost every f to this set is an isomorphism of $R \mid (F_0 \cup G_0)$ onto $R' \mid f(F_0 \cup G_0)$. Thus the sequences defined by the subsets F_i and G_i are (k, p)-identimorphic to one another. Since every finite subset H of E is contained in the k-th base of some restrictive sequence in $\mathscr{A}$, it follows from 4.2.2 that R is (k, p)-identimorphic to each restrictive sequence in $\mathscr{A}$. Thus the identity mapping on any given subset H is a (k, p)-isomorphism of R onto the sequence of restrictions to F_i $(i=0, 1,\ldots, k)$. For almost every mapping f and every r, the restriction of f to H is a (k, r)-isomorphism of the sequence $\{R \mid F_i\}$ onto $\{R' \mid f(F_i)\}$ (in fact, the restriction of almost every f to F_0 is an isomorphism of the first sequence onto the second). Finally, the identity mapping on $f(H)$ is by assumption a (k, p)-isomorphism of the second sequence onto R′. Thus, for almost every f, the restriction of f to H is a (k, p)-isomorphism of R onto R′.

Part 2 is now proved by letting k and p vary.

As to part 3, for any finite subset H of E′ and any k, p, the restriction of almost every f to E′ is the identity; hence the same holds for the restriction to H. This identity mapping is a (k, p)-isomorphism of R onto R′. Letting H, k, p vary, we see that R′ is a logical restriction of R. ◁

4.4.3. Theorem on logically equivalent extensions. *Let* R *and* S *be two multirelations with infinite bases, such that every finite restriction of* R *is isomorphic to a restriction of* S. *Then there exists an extension of*

R *which is logically equivalent to* S *and has the same cardinal as* R (see [HEN, 1953] and Exercise 5).

▷ There exists an injective projection filter $\mathscr{F}$ such that $R=\mathscr{F}^{-1}(S)$ (see 4.3.3). Let E be the base of R and E* a superset of E such that E* − E is equipollent to E. By 4.4.1, there exists an injective projection filter $\mathscr{F}^*$ extending $\mathscr{F}$ to E*. Moreover, for any finite subset H of E* and any k, p there exists a restrictive sequence of $R^*=(\mathscr{F}^*)^{-1}(S)$ whose k-th base is H, which is transformed by almost every mapping (mod $\mathscr{F}^*$) into a sequence which is (k, p)-identimorphic to S. It follows from 4.4.2 (2) that R* is logically equivalent to S. ◁

4.4.4.1. *Let* R *be a multirelation and* $\mathscr{A}$ *a logical or* δ*-logical class. If every finite restriction of* R *has an extension belonging to* $\mathscr{A}$, *then* R *has an extension in* $\mathscr{A}$.

▷ We present the argument for a logical class $\mathscr{A}$. Suppose that $\mathscr{A}$ is the union of two logical classes $\mathscr{A}'$ and $\mathscr{A}''$. We claim that the hypothesis of our assertion must hold either for $\mathscr{A}'$ or for $\mathscr{A}''$ (in place of $\mathscr{A}$). Indeed, otherwise there would exist two finite subsets F′ and F″ of the base such that $R \mid F'$ has no extension in $\mathscr{A}'$ and $R \mid F''$ no extension in $\mathscr{A}''$; thus $R \mid F' \cup F''$ would have no extension in $\mathscr{A}$. The argument remains valid if $\mathscr{A}$ is the union of finitely many logical classes. In particular, it is true for each $i=1, 2, \ldots$ when the classes in question are the (i, i)-equivalence classes $\mathscr{A}_i$; we then have $\mathscr{A}_{i+1} \subseteq \mathscr{A}_i$ for each i. By the compactness theorem (2.2.3), there exists a multirelation S in the intersection of the classes $\mathscr{A}_i$. For any subset F containing p elements of the base, the restriction $R \mid F$ has an extension in $\mathscr{A}_p$; this extension is (p, p)-equivalent to S, hence also $(1, p)$-equivalent to S. In other words, $R \mid F$ is a restriction of S (up to isomorphism). By 4.4.3, R has an extension which is logically equivalent to S and therefore belongs to $\mathscr{A}$. ◁

The above proof extends to the case of a δ-logical class (i.e., an intersection of logical classes): the classes $\mathscr{A}_i$ will be intersections of $\mathscr{A}$ with the (i, i)-equivalence classes (recall that the compactness theorem 2.2.3 remains valid for δ-logical classes).

4.4.4.2. Let R be a multirelation, F a finite subset of $|R|$, and k, p $(k \geqslant 1)$ natural numbers; let $\mathscr{A}$ be a logical or δ-logical class. Suppose that for every finite superset G of F there exists an extension S of $R \mid G$ which belongs to $\mathscr{A}$, such that the identity mapping on F is not a (k, p)-isomorphism of R onto S. Then R *has an extension* R′, *belonging to* $\mathscr{A}$, *such that the identity mapping on* F *is not a* $(k, p+h)$-*isomorphism of* R *onto* R′ (*where* $h = \text{card}\,\mathrm{F}$). Note that if F is empty this is simply the extension theorem 4.4.4.1.

▷ Denote the elements of F by a_i $(i=1,\ldots,h)$ and let A_i be the singleton relation of a_i (the unary relation equal to + for a_i, to − otherwise). Consider the δ-logical class consisting of the multirelations of the form $\mathrm{R}'\mathrm{A}'_1\ldots\mathrm{A}'_h$, where R′ is in $\mathscr{A}$, each A'_i $(i=1,\ldots,h)$ is the singleton of an element a'_i, the mapping taking each a_i to a'_i is a local isomorphism of $\mathrm{RA}_1\ldots\mathrm{A}_h$ onto $\mathrm{R}'\mathrm{A}'_1\ldots\mathrm{A}'_h$ but not a (k, p)-isomorphism. Then, using induction on k, one shows that the multirelations $\mathrm{RA}_1\ldots\mathrm{A}_h$ and $\mathrm{R}'\mathrm{A}'_1\ldots\mathrm{A}'_h$ are not $(k, p+h)$-equivalent; it then suffices to apply 4.4.4.1. ◁

4.4.5. Logical extension theorems. (1) *Let* R *be a multirelation with infinite base* E, *and* E* *a superset of* E *such that* E* − E *is at least equipollent to* E. *Then there exists a logical extension of* R *with base* E* (see [MAL, 1936], for a weaker version: the existence of a logically equivalent extension).

▷ Let $\mathscr{F}$ be a filter whose restriction to E is the identity mapping on E. By 4.4.1 with E′ = E, there exists an injective projection ultrafilter $\mathscr{F}^*$ extending $\mathscr{F}$ to E*. Moreover, for any finite subset H of E* and any k, p, there is a restrictive sequence of $\mathrm{R}^* = (\mathscr{F}^*)^{-1}(\mathrm{R})$ with k-th base H which is transformed by almost every mapping (mod $\mathscr{F}^*$) into a sequence which is (k, p)-identimorphic to R. It now follows from 4.4.2 (3) (with E′ and E replaced by E and E*, respectively) that R* is a logical extension of R. ◁

(2) *Let* S *be a multirelation and* R *a* 1-*extension of* S. *Then* R *has an extension which is a logical extension of* S *and has the same cardinal as* R.

▷ If the base |S| is finite, then R = S and the assertion is trivial. Suppose that the bases are infinite. There exists an injective |S|-identical projection filter such that $\mathrm{R} = \mathscr{F}^{-1}(\mathrm{S})$ (see 4.3.5). Let E* be a superset of the base of R such that E* − |R| is equipollent to |R|. By 4.4.1, there exists an injective projection ultrafilter $\mathscr{F}^*$ which is an extension of $\mathscr{F}$ to E* (and is therefore |S|-identical) satisfying the condition of 4.4.1. Set $\mathrm{R}^* = (\mathscr{F}^*)^{-1}(\mathrm{S})$. It now follows from 4.4.2(3) (with R, E, R′ replaced by R*, E*, S, respectively) that R* is a logical extension of S. ◁

4.4.6. COMPACTNESS THEOREM FOR EXTENSIONS (generalization of 2.2.3). *Let* R *be a multirelation and* $\mathscr{A}_1, \ldots, \mathscr{A}_i, \ldots$ *a (finite or infinite) sequence of logical or* δ*-logical classes whose intersection does not contain any extension of* R. *Then there is a finite subsequence of* $\mathscr{A}_i$ *whose intersection does not contain any extension of* R.

▷ We may assume without loss of generality that $\mathscr{A}_i \supseteq \mathscr{A}_{i+1}$ for each i. Suppose that each class $\mathscr{A}_i$ contains an extension of R. By adding classes to the sequence (if necessary) we may assume that for every logical class $\mathscr{B}$ all the $\mathscr{A}_i$ from some i on are either in $\mathscr{B}$ or in its complement.

Suppose that R has a finite base. Then the set of all extensions of R and their isomorphic images is a logical class (Volume 1, 5.7, referring to 4.8.3). This class contains all the $\mathscr{A}_i$ for sufficiently large i and, by the compactness theorem (2.2.3), the intersection of the entire sequence $\mathscr{A}_i$ is not empty and hence contains an extension of R.

Now let R have an infinite base E. By 2.2.3, there exists a multirelation S which is in all the classes $\mathscr{A}_i$, and the intersection of the $\mathscr{A}_i$ is the class of all multirelations logically equivalent to S. For each finite subset F of E, the class of extensions of R | F and its isomorphic images is a logical class, and it is not disjoint from any $\mathscr{A}_i$ since it contains every extension of R. Thus it contains all the $\mathscr{A}_i$ for sufficiently large i, and so contains S. By 4.4.3, R has an extension which is logically equivalent to S and is therefore in the intersection of the $\mathscr{A}_i$. ◁

4.4.7.1. *Let* $\mathscr{A}$ *be a* δ*-logical class. Then the class of restrictions of all multirelations in* $\mathscr{A}$ *is a* δ*-universal class* (i.e., an intersection of universal classes; see Volume 1, 4.8.7).

▷ The class $\mathscr{B}$ of restrictions of the multirelations in $\mathscr{A}$ is closed under isomorphism. Moreover, if R is a multirelation each of whose finite restrictions is in $\mathscr{B}$, then R is also in $\mathscr{B}$. Indeed, every finite restriction of R has an extension in $\mathscr{A}$; hence, by 4.4.4, R has an extension which is in $\mathscr{A}$. It follows that $\mathscr{B}$ is a δ-universal class (see Volume 1, 4.8.7). ◁

4.4.7.2. *Any logical class which is closed under passage to restrictions is a universal class.*

▷ By 4.4.7.1, the class $\mathscr{A}$ in question is an intersection of universal classes, say $\mathscr{A}_1, \ldots, \mathscr{A}_i, \ldots, \mathscr{A}_{i+1} \subseteq \mathscr{A}_i$ for each i. Since $\mathscr{A}$ is a logical class, the differences $\mathscr{A}_i - \mathscr{A}$ are logical classes with empty intersection. By the compactness theorem (2.2.3), we have $\mathscr{A}_i = \mathscr{A}$ for all sufficiently large i, so that $\mathscr{A}$ is a universal class. ◁

If $\mathscr{A}$ is a logical class, the δ-universal class of all restrictions of multirelations in $\mathscr{A}$ is not always a universal class. This may be seen from the logical class of groups: because of the Mal'tsev conditions, the class of all restrictions of groups is not a universal class.

4.4.8.1. *Let* R, S *be two logically equivalent multirelations. There exists a multirelation, logically equivalent to* R *and* S, *in which both* R *and* S *are embeddable. Moreover, we may assume that the cardinal of this multirelation is* max(card |R|, card |S|).

▷ By Volume 1, 3.2.6, with R in place of both M and N, and S in place of M′, there exists a multirelation T, in which R and S are embeddable, whose finite restrictions coincide with those of R (up to isomorphism), and whose base is the union of the bases of a restriction isomorphic to R and a restriction isomorphic to S. By 4.4.3, T has an extension which is logically equivalent to R (and S) and of the same cardinal as T. ◁

4.4.8.2. *Let* R, S *be two multirelations,* $\mathscr{A}$ *a logical or* δ*-logical class. Suppose that for any finite restrictions* R′, S′ *of* R, S, *respectively,* R′ *and* S′ *are embeddable in some multirelation of* $\mathscr{A}$. *Then* R *and* S *are embeddable in some multirelation of* $\mathscr{A}$. *Moreover, we may assume that the cardinal of this multirelation is* max(card |R|, card |S|).

▷ We present the proof for a logical class $\mathscr{A}$. Suppose it is the union of two disjoint logical classes $\mathscr{B}$, $\mathscr{C}$. Reasoning as in 4.4.4, we see that one of $\mathscr{B}$ or $\mathscr{C}$ may be substituted for $\mathscr{A}$ in the hypotheses of our proposition. For each natural number i, partition the logical classes into (i, i)-equivalence classes; we may now replace $\mathscr{A}$ by one of these classes $\mathscr{A}_i$ ($\mathscr{A}_{i+1} \subseteq \mathscr{A}_i$). By the compactness theorem (2.2.3) there exists a multirelation T which is a member of $\mathscr{A}_i$ for all i. For every finite restriction R′ of R and every i, there exists $T_i \in \mathscr{A}_i$ such that R′ is embeddable in T_i. Moreover, T_i is (i, i)-equivalent (and *a fortiori*, $(1, i)$-equivalent) to T: any restriction of T_i to at most i elements is embeddable in T. Let $i \geqslant \text{card}\,|R'|$; then it is clear that R′ is embeddable in T. Thus, any finite restriction of R is embeddable in T. By 4.4.3, there exists an extension R* of R which is logically equivalent to T and has the same cardinal as |R|. Similarly, there is an extension S^* of S, logically equivalent to T, of the same cardinal as |S|. By 4.4.8.1, there exists a multirelation P such that R^* and S^* (hence also R and S) are embeddable in P and, since P is logically equivalent to T, we have $P \in \mathscr{A}$. Moreover, we may assume that the cardinal of P is $\max(\text{card}\,|R^*|, \text{card}\,|S^*|) = \max(\text{card}\,|R|, \text{card}\,|S|)$. ◁

4.4.9. Let k be a natural number. Two multirelations are said to be (k, ω)-*equivalent* if they are (k, p)-equivalent for every natural p. By 1.3.3, this is equivalent to the statement that they assign the same value to any prenex formula in which there are at most k alternations of universal and existential quantifiers.

Let R, S *be* (k, ω)*-equivalent multirelations. Then there exists a multirelation in which both* R *and* S *are embeddable, and which is* (k, ω)*-equivalent to both. Moreover, we may assume that the cardinal of this multirelation is* $\max(\text{card}\,|R|, \text{card}\,|S|)$.

▷ This is obvious for $k=0$, for $(0, \omega)$-equivalence simply means that the multirelations have the same arity (and the same value in the case of 0-ary relations). If $k \geqslant 1$ our assertion follows from 4.4.8.2; indeed, R and S are $(1, \omega)$-equivalent and therefore have the same finite restrictions (up to isomorphism). ◁

4.5. Theorem on common logical extensions

If R *and* R′ *are logically equivalent, they possess isomorphic logical extensions* (assuming the axiom of choice).

Moreover, we may stipulate that *the cardinal of the isomorphic logical extensions is the maximum of the cardinals of* R *and* R′.

▷ Let us assume that the bases |R| and |R′| are denumerable; denote the elements of |R| by a_i $(i=0, 1, 2, \dots)$ and those of |R′| by a'_i.

Since R and R′ are logically equivalent, it follows that for each i, each finite subset F of |R| and each finite subset F′ of |R′|, there exists an (i, i)-isomorphism h of R onto R′ with domain a superset G of F and range a superset G′ of F′: $G = F \cup h^{-1}(F')$ and $G' = h(F) \cup F'$. For each i, we let

$$F_i = \{a_0, a_1, \dots, a_i\} \quad \text{and} \quad F'_i = \{a'_0, a'_1, \dots, a'_i\}$$

and let h_i be the (i, i)-isomorphism in question:

$$G_i = F_i \cup h_i^{-1}(F'_i) \quad \text{and} \quad G'_i = h_i(F_i) \cup F'_i.$$

Replacing the sequence of natural numbers $i=0, 1, 2, \ldots$ by a suitable infinite subsequence and renumbering the isomorphisms h_i and sets G_i, G'_i, we may assume that for each i and each $x \in F_i$ the images $h_j(x)$ ($j=i$, $i+1, i+2, \ldots$) are either all identical or all distinct. Moreover, in the latter case, each element $h_j(x)$ ($j=i, i+1, i+2, \ldots$) lies outside F'_j. Similarly, we may assume that for each i and each $x' \in F'_i$ the elements $h_j^{-1}(x')$ are either identical or all distinct, and in the latter case $h_j^{-1}(x') \notin F_j$ for each j. If the elements $h_j(x)$ are all identical, say to some x', we shall say that x and x' are *paired.*

For each i, we are going to define an injective mapping f_i of G_i into G_{i+1} whose restriction to F_i is the identity mapping; the restriction of the injective mapping $f'_i = h_{i+1} f_i h_i^{-1}$ of G'_i into G'_{i+1} to F'_i will then be the identity mapping. We set $f_i(x) = x$ for $x \in F_i$ and $f_i(x) = h_{i+1}^{-1} h_i(x)$ for $x \in G_i - F_i$. This is indeed a mapping of G_i into G_{i+1}: if $x \in F_i$ then $x \in F_{i+1}$, and so $x \in G_{i+1}$; while in the other case $x \in h_i^{-1}(F'_i)$, so that $h_i(x) \in F'_i$, $h_i(x) \in F'_{i+1}$, and finally $h_{i+1}^{-1} h_i(x) \in G_{i+1}$. Moreover, the mapping we have constructed is injective, for h_i and h_{i+1} are injective. Now, if $x \in F_i$ and $y \in G_i - F_i$, then $y \in h_i^{-1}(F'_i)$; hence there exists $k \leqslant i$ such that $y = h_i^{-1}(a'_k)$. Thus, either y and a'_k are paired, in which case $f_i(y) = y$ and $f_i(y) \neq f_i(x) = x$; or $h_{i+1}^{-1}(a'_k)$ is not in F_{i+1}, hence not in F_i, and so

$$f_i(y) = h_{i+1}^{-1} h_i(y) = h_{i+1}^{-1}(a'_k) \neq f_i(x) = x.$$

Instead of a single multirelation R, we now associate with each i two multirelations R_i and R'_i, isomorphic to R, so that each mapping f_i is the identity mapping on G_i; thus G_{i+1} becomes a superset of G_i for each i. Similarly, each mapping f'_i becomes the identity on G'_i. Now, for each i and each $j \leqslant i$ we consider the sequences of j elements of the base $|R_i|$. Let us say that any two such sequences are i-equivalent if the mapping defined as the identity on G_i and taking each term of the one sequence onto the similarly positioned term of the other is an $(i, i-j)$-automorphism of R_i. Then, for any fixed i and $j=0, 1, \ldots, i$, there are only finitely many i-equivalence classes (see 1.2.1). We now take a representative sequence of each class. Beginning with G_0 and going through the natural number sequence, we replace each set G_{i+1} by the first set G_j ($j=i+1, i+2, \ldots$) which contains the elements of all the representative sequences. We do the same for the sets G'_i. What we have done amounts to replacing the natural numbers $i=0, 1, 2, \ldots$ by an infinite subsequence and renumbering the h_i, G_i, G'_i in such a way that the conditions stipulated above remain valid.

We now show that we may assume that, for each i, the identity mapping on G_i is an (i, i)-isomorphism of R_i onto R_{i+1}, and similarly that the identity mapping on G'_i is an (i, i)-isomorphism of R'_i onto R'_{i+1}. For each i, partition the natural numbers into i-equivalence classes, declaring two numbers j and k ($j \leqslant k$) to be i-equivalent if the identity mapping on G_j is an (i, i)-isomorphism of R_j onto R_k. Since the number of i-equivalence classes is finite, we may replace the sequence of natural numbers by an infinite subsequence, consisting of all the 0-equivalent numbers; then, retaining the first term, we replace the remainder by a sequence of 1-equivalent numbers, and so on.

Let G be the union of the sets G_i and S the relation with base G which is a common extension of the relations $R \mid G_i$. Note that S is an extension of R. Indeed, the base of R is the union of the sets F_i, each of which is a subset of G_i. Moreover, the restrictions $R \mid F_i$ and $R_i \mid F_i$ are identical for each i, since the isomorphism of R onto R_i was introduced specifically in order to make f_i act as the identity on G_i, while f_i left the elements of F_i fixed from the start. Thus the identity mapping on F_i may be extended to an isomorphism of R onto R_i; it is therefore a (k, p)-isomorphism of R onto R_i for all k, p. On the other hand, it follows from 2.1.1 that the identity mapping on G_i, hence also on F_i, is an (i, i)-isomorphism of R_i onto S. Finally, the identity mapping on F_i is an (i, i)-isomorphism of R onto S. Varying i and observing that every finite subset of $|R|$ is contained in F_i for all sufficiently large i, we see that S is a logical extension of R. Similarly we define a relation S′ on the union G′ of the sets G'_i and show that S′ is a logical extension of R′. Finally, the union of the local isomorphisms h_i yields an isomorphism of S onto S′. ◁

The above proof is due to J. L. Paillet.

4.6. Logical morphism and logical embedding

We now consider a (k, p)-isomorphism of a special type; for a general account, the reader may consult [BEN, 1970]. Any local isomorphism of R onto S will be called a *$(0, p)$-morphism* for any p. Let F be the domain of the local isomorphism f; we shall say that f is a *(k, p)-morphism* of R onto S ($k \geqslant 1$) if, for any $q \leqslant p$ and any set $\bar{F}$ obtained by adding q elements of |R| to F, there exists an extension $\bar{f}$ of f to $\bar{F}$ which is a $(k-1, p-q)$-morphism of R onto S. We shall say that R is *(k, p)-embeddable* in S if the empty mapping is a (k, p)-morphism of R onto S. (k, p)-embedding is reflexive and transitive.

R is $(1, p)$-embeddable in S if every restriction of R to at most p elements is isomorphically embeddable in S.

We shall say that R is *logically embeddable* in S if it is (k, p)-embeddable in S for all k, p. In particular, this is the case when R is isomorphically embeddable in S or logically equivalent to S.

R *is logically embeddable in* S *if and only if every finite restriction of* R *is isomorphically embeddable in* S.

▷ If R is logically embeddable in S, it is in particular $(1, p)$-embeddable in S for any p. Thus any restriction to a set of finite cardinality p is (isomorphically) embeddable in S. Conversely, if this condition is satisfied, it follows from 4.4.3 that R has an extension T which is logically equivalent to S. Thus R is logically embeddable in T, hence also in S. ◁

Let S *be an extension of* R. *Then* S *is a* 1*-extension of* R *if and only if, for any finite subset* F *of* |R| *and any* k, p, *the identity mapping on* F *is a* (k, p)*-morphism of* S *onto* R.

▷ If the condition is satisfied, the identity mapping on F is a $(1, p)$-morphism of S onto R for any p; hence it is a $(1, p)$-isomorphism, so that S is a 1-extension of R (see 4.3.5). Conversely, if S is a 1-extension of R, it follows from 4.4.5.2 that S has an extension T which is a logical extension of R. Then, for any finite subset F of |R| and any k, p, the identity mapping on F is a (k, p)-isomorphism of T onto R and also of S onto T, hence of S onto R. ◁

EXERCISES

1

A multirelation R with infinite base E is said to be *discrete* if it is the logical limit of a sequence of multirelations with finite bases (any logical class containing R contains all terms of the sequence from some term on; see 2.1).

(1) Show that if R is a discrete chain in the above sense it is discrete in the ordinary sense: each element has a predecessor and a successor, except the minimal and maximal elements, which exist; show that the converse is also true.

(2) Show that for any multirelation R there exists a discrete multirelation with the same age as R (i.e., with the same finite restrictions, up to isomorphism; see Volume 1, 3.2.4). *Hint:* Consider a sequence of finite restrictions R_i $(i=1,2,\ldots)$ of R such that every finite restriction of R is embeddable in R_i for all sufficiently large i. Then take a suitable logically convergent subsequence (see 2.2.2).

(3) Show that for any multirelation R there exists a discrete multirelation which is an extension of R and has the same age. *Hint:* Let S be a discrete multirelation of the same age as R, and use the extension theorem 4.4.3.

(4) Let p be a prime and G a finite group with identity element e. Prove that either there exists an element $x \neq e$ such that $x^p = e$, or for every $x \in G$ there is an element $y \in G$ such that $y^p = x$. (*Hint:* let q be such that $x^q = e$: then q is not divisible by p and there exist integers i, j such that $pi + qj = 1$; thus $x^{pi}x^{qj} = x$ and since $x^{qj} = e$ this gives $(x^i)^p = x$.) For each prime p, construct a bound formula taking the value + for all finite groups and − for the multiplicative group of rationals. Hence deduce that the latter is not a discrete relation (communicated by A. Blanchard and M. H. Dulac).

2

Let R and S be two multirelations. Prove the following refinement of 4.4.3:

Suppose that for every finite subset F *of the base* |R| *and any natural number p there exists a* (1, *p*)*-isomorphism of* R *onto* S *with domain* F. *Then there exists a* 1*-extension of* R *which is logically equivalent to* S.

By assumption, we may associate with every pair (F, p) the set $U_{F,p}$ of mappings f of |R| into |S| such that the restriction of f to F is a $(1, p)$-isomorphism of R onto S. The sets $U_{F,p}$ and all their supersets constitute an injective projection filter, and R is the inverse projection of S relative to this filter; let $\mathscr{F}$ be a refinement of the filter which is an ultrafilter.

(1) Suppose that $E = |R|$ is infinite and let E^* be a superset of E. Let $\mathscr{F}^*$ be an injective projection filter of E^* onto |S| which extends $\mathscr{F}$ (see 4.3.6). Show that the inverse projection $(\mathscr{F}^*)^{-1}(S)$ is a 1-extension of R.

(2) Suppose that the cardinal of $E^* - E$ is at least that of E. Using 4.4.1 and 4.4.2, prove the desired assertion.

(3) The assumption of our assertion is not symmetric with respect to R and S. Show this by letting R be the chain of natural numbers and $S = R + 1$.

3

(1) A logical formula is said to be an $\forall\exists$*-formula* if it is prenex and its prefix consists of universal quantifiers followed by existential quantifiers; the corresponding logical class is known as an $\forall\exists$*-class*. The definitions of $\exists\forall$*-formula* and $\exists\forall$*-class* are analogous.

Let R and S be multirelations of the same arity. Show that the following three conditions are equivalent:

(i) every $\exists\forall$-class containing R also contains S;
(ii) every $\forall\exists$-class containing S also contains R;
(iii) for any finite subset F of $|R|$ and any natural number p, there exists a $(1, p)$-isomorphism of R onto S with domain F.

Using the preceding exercise, show that this condition is not symmetric with respect to R and S.

(2) Consider a family of multirelations R_i which is totally ordered by extension: R_j is an extension of R_i whenever $j>i$. Define the *limit* of the multirelations R_i as their common extension to the union of their bases.

We wish to establish the following theorem ([ŁOS-SUS, 1957] and [CHA, 1959]):

A δ-logical class is an intersection of $\forall\exists$-classes if and only if it is closed under passage to limits.

Necessity is obvious; we therefore prove sufficiency. Let $\mathscr{A}$ be a δ-logical class and $\mathscr{A}^*$ the intersection of the $\forall\exists$-classes containing $\mathscr{A}$. We shall show that $\mathscr{A}^*=\mathscr{A}$ if $\mathscr{A}$ is closed under passage to limits.

By the compactness theorem 2.2.3, for every multirelation R in $\mathscr{A}^*$ there exists a multirelation S in $\mathscr{A}$ such that every $\exists\forall$-class containing R also contains S. Using the preceding exercise, show that there exists a 1-extension of R which is logically equivalent to S and is therefore in $\mathscr{A}$.

Beginning with $R_1=R$, which is in $\mathscr{A}^*$, consider a 1-extension S_1 of R_1 which belongs to $\mathscr{A}$. Then take an extension R_2 of S_1 which is a logical extension of R_1 (see 4.4.5(2)). Iterating the procedure, we obtain two sequences $R_1, \ldots, R_i, \ldots$ and $S_1, \ldots, S_i, \ldots$, where each S_i is a 1-extension of R_i which belongs to $\mathscr{A}$, and each R_{i+1} is an extension of S_i and a logical extension of R_i. The common limit of both sequences is in $\mathscr{A}$ if $\mathscr{A}$ is closed under passage to limits. On the other hand, this common extension is a logical extension of $R_1=R$ by 1.4.1(3); thus R belongs to $\mathscr{A}$ (this method is due to [SHO, 1967]).

(3) It is essential that the class $\mathscr{A}$ be δ-logical. Indeed, let C be the succession relation and Z the singleton of zero over the natural numbers; then (C, Z) and all isomorphic birelations form a class which is closed under passage to limits but not under logical equivalence (this remark is due to A. Roberty).

4

Every embedding logical class (Boolean combination of universal classes; see Volume 1, 5.7) is both an $\forall\exists$-class and an $\exists\forall$-class. To be precise: an embedding class is representable by a permuting free formula (Volume 1, Chapter 5, Exercise 6) preceded by quantifiers in arbitrary order; prove the converse.

5

We wish to generalize the concept of *relation*, considering a set I whose elements will be called *indices*. Given some finite set $\mu\subseteq I$, a *μ-ary relation* (or relation of arity μ) *with base* E will associate a value + or − with each sequence of elements of E, indexed by the elements of μ. An *(m, μ)-ary operator*, where m is a natural number, transforms each m-ary relation (in the usual sense) into a μ-ary relation with the same base. The concepts of restriction, isomorphism, local isomorphism and free operator are generalized in the natural way.

(1) Generalization of the *compactness lemma* for free operators (Volume 1, 4.4). We start with the usual m-ary relations; for each k, we have a free operator $\mathscr{P}_k$ which transforms each m-ary relation R with base E into a relation of arity $\mu_k\subseteq I$. A given sequence $x_i, \ldots, x_j$ of elements of E, indexed by elements $i, \ldots, j$ of μ_k, determines a value $\mathscr{P}_k(R)\,(x_i, \ldots, x_j)=+$

or $-$. We may clearly consider sequences indexed by *all* the elements of I, only the elements with indices in μ_k being active. *Compactness lemma:* If every finite subset of the given family of operators is consistent, then the entire family is consistent (i.e., takes the value $+$ for a certain m-ary relation and a certain sequence of base elements indexed by I). For the proof, note that we may identify each element x_i of the base with its index i by defining an equivalence relation on I which we call *identification*: $i, j \in \mathrm{I}$ are equivalent if $x_i = x_j$. We then replace R by a birelation (R, U), where U is "identification" and R retains the same value whenever any indices are replaced by U-equivalent indices. With each finite set K of operators we associate the set $\mathrm{U_K}$ of birelations with base I which give these operators the value $+$; by assumption, none of the sets $\mathrm{U_K}$ is empty, and if $\mathrm{K}' \supseteq \mathrm{K}$ then $\mathrm{U_{K'}} \subseteq \mathrm{U_K}$. Thus the family of all supersets of the sets $\mathrm{U_K}$ constitutes a filter over the above birelations. Now consider an ultrafilter refining this filter. Note that, for any finite set $\mu \subseteq \mathrm{I}$, if we partition the birelations into finitely many equivalence classes, each corresponding to a given restriction on μ, exactly one of these classes is an element of the ultrafilter. The common extension of the birelations is a birelation with base I which gives all the operators in question the value $+$.

(2) Generalization of the *compactness theorem* (see 2.2.3). We define a *logical formula* in the generalized sense to be a formula built up as usual from connections, quantifiers and generalized free operators. We may assume that the formulas are *prenex* and that the *bound* indices (in the scope of quantifiers) are natural numbers; the free indices, however, are elements of the set I. The truth value $+$ or $-$ of a formula for an m-ary relation and a sequence of elements indexed by I is defined in the natural way; the same holds for the value of a birelation consisting of an identification on I and a compatible m-ary relation. *Compactness theorem:* If every finite subset of a given family of formulas is consistent, then the entire family is consistent. To prove this, we enrich the family of formulas as follows, using a technique due to Kalmár and Henkin. For each formula beginning with $\forall$, we add all formulas obtained by omitting the quantifier and replacing its index (a natural number) by each index of I in turn; for each formula beginning with $\exists$, we add a formula obtained by omitting this quantifier and replacing its index by a new element, which is added to I (unless the family already contains a formula of this type). Show that the resulting family of formulas is again such that each finite subset is consistent and the family is closed under the two operations just described. Apply the compactness lemma to the new family of free formulas, and finally show that the original formulas are satisfied by the same systems (m-ary relation on I and identification).

(3) Prove the *extension theorems* of 4.4 as follows. Given a relation R with base D and a logical class $\mathscr{A}$, if every restriction of R to a finite subset F of D has an extension in $\mathscr{A}$, then the finite set consisting of all free formulas with indices in F which take the value $+$ for R and the sequence of their indices, and in addition a bound formula A representing $\mathscr{A}$, is consistent. By the compactness theorem, the set consisting of all free formulas with indices in D which take the value $+$ for R and the sequence of their indices, and in addition the bound formula A, is consistent. Hence deduce that R has an extension satisfying A. Repeat this proof for the case that every finite restriction of R has an extension which is logically equivalent to a given relation S; in this case, instead of the single bound formula A, one considers the set of all bound formulas which are true for S.

A full account of the subject of this exercise may be found in [HEN, 1949 and 1953].

CHAPTER 5

THEORIES AND AXIOM SYSTEMS

5.1. THEORY; CONSISTENCY; INTERSECTION OF THEORIES

5.1.1. Let $\mathcal{M}$ be a class of multirelations of the same arity μ. The set $\mathcal{T}$ of bound formulas of predicarity μ satisfied by every multirelation in $\mathcal{M}$ (see Volume 1, 5.4) will be called the *theory associated with* $\mathcal{M}$; we shall call μ the arity of $\mathcal{T}$.

It is evident that $\mathcal{T}$ is closed under deduction and conjunction: if $\mathrm{P} \in \mathcal{T}$ and $\mathrm{P} \vdash \mathrm{Q}$, then $\mathrm{Q} \in \mathcal{T}$ (where P and Q are bound formulas); if $\mathrm{P} \in \mathcal{T}$ and $\mathrm{Q} \in \mathcal{T}$, then $\mathrm{P} \bigwedge \mathrm{Q} \in \mathcal{T}$. Any bound thesis of predicarity μ belongs to $\mathcal{T}$. Finally, if $\mathrm{P} \in \mathcal{T}$ and $\mathrm{P} \Rightarrow \mathrm{Q} \in \mathcal{T}$, then $\mathrm{P} \bigwedge (\mathrm{P} \Rightarrow \mathrm{Q}) \in \mathcal{T}$; since $\mathrm{P} \bigwedge (\mathrm{P} \Rightarrow \mathrm{Q}) \vdash \mathrm{Q}$, it follows that $\mathrm{Q} \in \mathcal{T}$.

If $\mathcal{M}$, $\mathcal{M}'$ are two classes of multirelations of arity μ, and $\mathcal{T}$, $\mathcal{T}'$ are the associated theories, then $\mathcal{M} \subseteq \mathcal{M}'$ implies $\mathcal{T} \supseteq \mathcal{T}'$. Instead of saying that $\mathcal{T}'$ is contained in $\mathcal{T}$, we shall sometimes say that $\mathcal{T}'$ is *weaker* than $\mathcal{T}$, or that $\mathcal{T}$ is *stronger* than $\mathcal{T}'$.

It will be convenient to include the case that $\mathcal{M}$ is empty; the associated theory will then be the *set of all bound formulas* of predicarity μ. This theory is stronger than any other theory of arity μ. Moreover, it is the only theory that contains antitheses, and indeed contains formulas together with their negations; we shall call it the *inconsistent theory* of arity μ. Any other theory is said to be *consistent*.

If $\mathcal{M}$ is the class of all μ-ary multirelations, $\mathcal{T}$ is simply the set of bound theses of predicarity μ; this is the weakest theory, and it is consistent.

5.1.2. We now present a few examples of consistent theories which do not reduce to the set of theses.

(1) Let $\mathcal{M}$ be the class of all sets with at least p elements (where p is a positive integer); the associated theory is the set of all formulas deducible from

$$\mathrm{A}_p = \underset{1,2,\ldots,p}{\exists} x^1 \not\equiv x^2 \bigwedge \cdots \bigwedge x^1 \not\equiv x^p \bigwedge x^2 \not\equiv x^3 \bigwedge \cdots \bigwedge x^{p-1} \not\equiv x^p.$$

Let $\mathcal{M}$ be the set of all sets with at most p elements; the associated theory is the set of all formulas deducible from $\neg A_{p+1}$. The theory associated with the class of sets of p elements is the set of formulas deducible from $A_p \wedge \neg A_{p+1}$.

Let $\mathcal{M}$ be the class of infinite sets; the associated theory is the set of formulas A_p, $p=1, 2, 3, \ldots$ and all formulas deducible from them (note that $A_{p+1} \vdash A_p$ for every p).

(2) The theory associated with the class of reflexive binary relations is the set of all formulas deducible from $\underset{x}{\forall} \rho xx$. The theory associated with the class of pre-orderings (reflexive transitive relations) is the set of all formulas deducible from the conjunction of $\underset{x}{\forall} \rho xx$ and $\underset{xyz}{\forall} (\rho xy \wedge \rho yz) \Rightarrow \Rightarrow \rho xz$. Similarly, for the class of orderings we take the conjunction of the preceding formulas with $\underset{xy}{\forall} (\rho xy \wedge \rho yx) \Rightarrow x \equiv y$. The theory associated with the class of chains (total orderings) is generated by the conjunction of the preceding formulas with $\underset{xy}{\forall} \rho xy \vee \rho yx$. To obtain the class of chains with a minimal element such that every element has a successor, we take the conjunction of all the previous formulas with the formulas $\underset{x}{\exists} \underset{y}{\forall} \rho xy$ and

$$\underset{x}{\forall} \underset{y}{\exists} x \not\equiv y \wedge \rho xy \wedge \underset{z}{\forall} (\rho zx \vee \rho yz).$$

The theory associated with the class of wellorderings (chains in which every restriction has a minimal element) is strictly stronger than any of the preceding theories. For example, it contains a formula which is true for all chains in which the set of elements of the second kind (elements with no predecessor) contains a minimal element.

(3) The theory associated with the class of groups (viewed as a class of ternary relations) may be called the *theory of groups* (not to be confused with the "theory of groups" in the conventional sense, which is a far more complicated theory, containing representations of the concepts of subgroup and the set of elements of a group). Our theory of groups is the set of all formulas deducible from the following three formulas:

$$\underset{xy}{\forall} \underset{z}{\exists} \sigma xyz \wedge \underset{xyzz'}{\forall} (\sigma xyz \wedge \sigma xyz') \Rightarrow z \equiv z'$$

(characterizing functional ternary relations)

$$\forall_{xyztuv} (\sigma xyt \bigwedge \sigma tzu \bigwedge \sigma yzv) \Rightarrow \sigma xvu$$

(characterizing associative relations)

$$\exists_{u} \left(\forall_{x} \sigma xux \bigwedge \sigma uxx \bigwedge \forall_{x} \exists_{y} \sigma xyu \right)$$

(characterizing relations with a neutral element such that each element has an inverse).

The theory of abelian groups is generated by the conjunction of the above three formulas and the formula

$$\forall_{xyz} \sigma xyz \Rightarrow \sigma yxz.$$

5.1.3. Application of the compactness theorem; syntactic characterization of theories

(1) *Any set of bound formulas closed under deduction and conjunction is a theory* (and conversely, by 5.1.1).

▷ With each bounded formula we associate the logical class that it represents. Then any set of the type mentioned in the assertion, $\mathcal{T}$ say, becomes a set of logical classes, which is closed under inclusion in the sense that if $\mathcal{M} \in \mathcal{T}$ and $\mathcal{M}' \supseteq \mathcal{M}$ then $\mathcal{M}' \in \mathcal{T}$, and closed under intersection: if $\mathcal{M} \in \mathcal{T}$ and $\mathcal{M}' \in \mathcal{T}$ then $\mathcal{M} \cap \mathcal{M}' \in \mathcal{T}$. Let $\mathcal{M}_0$ be the intersection of all classes $\mathcal{M} \in \mathcal{T}$; we claim that $\mathcal{T}$ is the theory associated with $\mathcal{M}_0$. In other words, we must show that if $\mathcal{A}$ is a logical class such that $\mathcal{A} \supseteq \mathcal{M}_0$, then $\mathcal{A} \in \mathcal{T}$.

Let $\mathcal{A}'$ be the complement of $\mathcal{A}$; since $\mathcal{A} \supseteq \mathcal{M}_0$, the intersection of $\mathcal{A}'$ and the classes $\mathcal{M} \in \mathcal{T}$ is empty. But $\mathcal{A}'$ and the $\mathcal{M} \in \mathcal{T}$ are logical classes, and so it follows from the compactness theorem (2.2.3) that there exist finitely many classes $\mathcal{M}_1, \ldots, \mathcal{M}_h \in \mathcal{T}$ such that $\mathcal{M}_1 \cap \cdots \cap \mathcal{M}_h \cap \mathcal{A}'$ is empty. Thus $\mathcal{M}_1 \cap \cdots \cap \mathcal{M}_h \subseteq \mathcal{A}$. Since $\mathcal{T}$ is closed under intersection and inclusion, it follows that $\mathcal{A} \in \mathcal{T}$. ◁

It is not sufficient to consider a set of bound formulas closed under deduction. For example, if P and $\neg$P are consistent bound formulas, the set of formulas deducible from P or from $\neg$P is closed under deduction but is not a theory – if it were, it would contain the conjunction P$\bigwedge\neg$P.

(2) *Let $\mathscr{T}$ be a set of bound formulas which contains every thesis and has the property: if* $P \in \mathscr{T}$ *and* $P \Rightarrow Q \in \mathscr{T}$, *then* $Q \in \mathscr{T}$, *where* P *and* Q *are arbitrary bound formulas. Then* $\mathscr{T}$ *is a theory* (and conversely, by 5.1.1).

▷ $\mathscr{T}$ is closed under deduction, for if $P \in \mathscr{T}$ and $P \vdash Q$, then $P \Rightarrow Q$ is a thesis, so that $P \Rightarrow Q \in \mathscr{T}$ and therefore $Q \in \mathscr{T}$. Moreover, $\mathscr{T}$ is closed under conjunction, for if $P \in \mathscr{T}$ and $Q \in \mathscr{T}$, the fact that $Q \vdash P \Rightarrow (P \wedge Q)$ implies that

$$P \Rightarrow (P \wedge Q) \in \mathscr{T} \quad \text{and} \quad P \wedge Q \in \mathscr{T}. \ \triangleleft$$

5.1.4. Any intersection of theories is a theory. In fact, if the given theories $\mathscr{T}_i$ are associated with classes $\mathscr{M}_i$, then their intersection $\bigcap \mathscr{T}_i$ is associated with the union $\bigcup \mathscr{M}_i$.

On the other hand, the set-theoretic union of two theories need not be a theory. This is demonstrated by the previously given example of two consistent bound formulas P and $\neg P$; the formulas deducible from P form a theory, as do those deducible from $\neg P$, but the union of these two theories is not a theory.

5.2. Axiom system, class of models; union-theory, finitely-axiomatizable theory, saturated theory

Let $\mathscr{U}$ be a set of bound formulas of predicarity μ. A *model* of $\mathscr{U}$ is a μ-ary multirelation which satisfies every formula of $\mathscr{U}$, in other words, a multirelation which is a model of every formula of $\mathscr{U}$ (see Volume 1, 5.4).

Let $\overline{\mathscr{U}}$ be the theory associated with the class $\mathscr{M}$ of models of $\mathscr{U}$; $\overline{\mathscr{U}}$ is the intersection of all theories containing $\mathscr{U}$. It is known as the theory *generated* by $\mathscr{U}$; alternatively, we shall say that $\mathscr{U}$ is an *axiom system* for the theory $\overline{\mathscr{U}}$.

Clearly, $\overline{\mathscr{U}} \supseteq \mathscr{U}$. If $\mathscr{U}$ is a theory, then $\overline{\mathscr{U}} = \mathscr{U}$, and conversely. Thus, for any set $\mathscr{U}$ of bound formulas, we have $\overline{\overline{\mathscr{U}}} = \overline{\mathscr{U}}$. Finally, if $\mathscr{U} \subseteq \mathscr{V}$, where $\mathscr{V}$ is a set of bound formulas of predicarity μ, then the class of models of $\mathscr{V}$ is contained in the class of models of $\mathscr{U}$, so that $\overline{\mathscr{U}} \subseteq \overline{\mathscr{V}}$.

We shall adopt the convention that every multirelation is a model of the empty set of formulas. Thus the theory generated by the empty axiom system is the set of bound theses of predicarity μ. The corresponding class of models is the class of all μ-ary multirelations.

Examples of axiom systems: An axiom system for the theory consisting of all formulas deducible from a formula P is the single formula P. An axiom system for the theory associated with the class of infinite sets is the set of formulas A_p, $p=1, 2, \ldots$ (see 5.1.2(1)).

5.2.1. *The theory generated by a set of formulas* $\mathscr{U}$ *is the set of all bound formulas* Q *for which there exist finitely many formulas* $P_1, \ldots, P_h$ *in* $\mathscr{U}$ *such that* $P_1 \wedge \cdots \wedge P_h \vdash Q$.

$\triangleright$ On the one hand, $P_1 \wedge \cdots \wedge P_h$ belongs to any theory containing $P_1, \ldots, P_h$, and hence so does Q. Thus Q is in the theory $\overline{\mathscr{U}}$ generated by $\mathscr{U}$. This holds for all formulas Q of the type specified. On the other hand, the set of these formulas Q is closed under deduction and conjunction, and so it is a theory containing $\mathscr{U}$; thus the set of Q's contains $\overline{\mathscr{U}}$. $\triangleleft$

5.2.2. Class of models generated by a class of multirelations. With any class $\mathscr{M}$ of multirelations of arity μ we may associate the intersection $\overline{\mathscr{M}}$ of all logical classes containing $\mathscr{M}$, or, equivalently, the class of logical limits of all convergent sequences of members of $\mathscr{M}$ (see 2.1). It is evident that $\overline{\mathscr{M}} \supseteq \mathscr{M}$ and $\overline{\overline{\mathscr{M}}} = \overline{\mathscr{M}}$ If $\mathscr{M}'$ is another class of arity μ and $\mathscr{M} \subseteq \mathscr{M}'$, then $\overline{\mathscr{M}} \subseteq \overline{\mathscr{M}'}$. Moreover, $\mathscr{M} = \overline{\mathscr{M}}$ if and only if $\mathscr{M}$ is an intersection of logical classes.

For any $\mathscr{M}$, the class $\overline{\mathscr{M}}$ is closed with respect to logical equivalence. The converse, however, is false: the class $\mathscr{M}$ of finite sets is closed with respect to logical equivalence but is not an intersection of logical classes. In fact, we known that the only logical class containing $\mathscr{M}$ is the class of all (finite and infinite) sets (see Volume 1, 5.6.3); in this case, therefore, $\overline{\mathscr{M}}$ consists of all sets. Similarly, if $\mathscr{M}$ is the class of finite sets of even cardinality, then $\overline{\mathscr{M}}$ is the union of $\mathscr{M}$ and the class of all infinite sets.

For any class $\mathscr{M}$, the associated theory $\mathscr{T}$ is the set of all formulas representing logical classes containing $\mathscr{M}$. Thus the intersection $\overline{\mathscr{M}}$ of these classes is precisely *the class of all models* of $\mathscr{T}$.

If a class of models contains multirelations of arbitrarily large finite cardinality, then it contains infinite multirelations (see 2.2.4).

5.2.3. For any given theories $\mathscr{T}_i$, the intersection of all theories that contain $\bigcup_i \mathscr{T}_i$ will be called their *union-theory*. The union-theory is also the theory generated by $\bigcup_i \mathscr{T}_i$.

Given two formulas P and Q, the union-theory of the set of formulas deducible from P and the set of formulas deducible from Q is the set of formulas deducible from $P \wedge Q$.

Consider the theory associated with the class of sets with at least p elements (the formulas deducible from A_p; see 5.1.2(1)). The union-theory of these theories, $p=1, 2, \ldots$, is the theory associated with the class of infinite sets.

A set of theories is said to be *compatible* or *incompatible* according as their union-theory is consistent or inconsistent.

Let $\mathcal{M}_i$ be classes of multirelations and $\mathcal{T}_i$ the associated theories. The theory associated with $\bigcap_i \mathcal{M}_i$ is stronger than the union-theory of the $\mathcal{T}_i$. It may be strictly stronger: let $\mathcal{M}_1$ consist of a single relation R_1 and $\mathcal{M}_2$ of a relation R_2 distinct from but logically equivalent to R_2. The intersection of these classes is empty, so that the associated theory is the inconsistent theory; however, the theory associated with each of these classes is the same, $\mathcal{T}$ say, and the union-theory is of course $\mathcal{T}$ itself.

Nevertheless, if each class $\mathcal{M}_i$ is the class of models of $\mathcal{T}_i$, then the theory associated with $\bigcap_i \mathcal{M}_i$ is the union-theory of the theories $\mathcal{T}_i$.

5.2.4. A theory $\mathcal{T}$ is said to be *finitely-axiomatizable* if it has an axiom system consisting of finitely many formulas or, equivalently, of a single formula (the conjunction of all the "axioms"). Another, equivalent definition is that the class of models of $\mathcal{T}$ is a logical class.

Like the logical classes, the finitely-axiomatizable theories constitutes a denumerably infinite set. An example of a non-finitely-axiomatizable theory is the theory of infinite sets, generated by the formulas A_p of 5.1.2. In fact, by Volume 1, 5.5.2, any logical class containing the infinite sets must contain all finite sets of sufficiently large cardinality, so that the set of formulas A_p cannot be replaced by a finite set. Other examples were given in 1.5, 3.4.3 and 3.5.3.

5.2.5. If $\mathcal{U}$ and $\mathcal{V}$ are sets of bound formulas of the same predicarity, then

$$\overline{\mathcal{U} \cap \mathcal{V}} \subseteq \overline{\mathcal{U}} \cap \overline{\mathcal{V}} \quad \text{and} \quad \overline{\mathcal{U} \cup \mathcal{V}} \supseteq \overline{\mathcal{U}} \cup \overline{\mathcal{V}}.$$

Indeed, $\mathcal{U} \cap \mathcal{V} \subseteq \mathcal{U}$, and therefore $\overline{\mathcal{U} \cap \mathcal{V}} \subseteq \overline{\mathcal{U}}$; similarly, $\overline{\mathcal{U} \cap \mathcal{V}} \subseteq \overline{\mathcal{V}}$. A similar argument verifies the second formula (for $\cup$). It may happen that

$$\overline{\mathcal{U}\cap\mathcal{V}}\neq\overline{\mathcal{U}}\cap\overline{\mathcal{V}};$$

To see this, let P and Q be consistent formulas which are not theses, such that $P\vdash Q$ (if equideducible formulas are identified, we must assume that Q is strictly deducible from P). Now set $\mathcal{U}=\{P\}$ and $\mathcal{V}=\{Q\}$. Then $\overline{\mathcal{U}}\cap\overline{\mathcal{V}}$ is the theory generated by Q, but $\mathcal{U}\cap\mathcal{V}$ is empty, so that $\overline{\mathcal{U}\cap\mathcal{V}}$ is the set of theses. Similarly, it may happen that $\overline{\mathcal{U}\cup\mathcal{V}}\neq\overline{\mathcal{U}}\cup\overline{\mathcal{V}}$. Indeed, let P be a consistent formula which is not a thesis, and set $\mathcal{U}=\{P\}$ and $\mathcal{V}=\{\neg P\}$. Then $\overline{\mathcal{U}\cup\mathcal{V}}$ is the inconsistent theory, while $\overline{\mathcal{U}}\cup\overline{\mathcal{V}}$ is the set of all formulas deducible either from P or from $\neg P$, which, as we have seen, is not a theory.

5.2.6. If we consider a class consisting of a single multirelation M, the associated theory $\mathcal{T}$ is the set of all formulas true for M, which is known as the *saturated theory* of M. For any bound formula P to which M is assignable, we have either $P\in\mathcal{T}$ or $\neg P\in\mathcal{T}$, according as $P(M)=+$ or $P(M)=-$. Conversely, if the latter condition is valid, subject to the understanding that *only* one of the alternatives $P\in\mathcal{T}$ or $\neg P\in\mathcal{T}$ can hold, then $\mathcal{T}$ is saturated. The class of models of $\mathcal{T}$ consists of all multirelations logically equivalent to M. The only theory strictly stronger than a saturated theory is the inconsistent theory.

For any consistent theory $\mathcal{T}$, there is at least one saturated theory stronger than $\mathcal{T}$, namely, the theory defined by any model of $\mathcal{T}$. Moreover, $\mathcal{T}$ is the intersection of all saturated theories stronger than $\mathcal{T}$.

A theory may be both saturated and finitely-axiomatizable. In other words, a logical class may be a logical equivalence class. We thus get the logical equivalence class of a finitely-axiomatizable multirelation (see Volume 1, 7.4). For examples of saturated theories which are not finitely-axiomatizable, see 1.6 and 3.5.3.

5.3. Complement of a theory ([TAR, 1930]; see also [ŁOS, 1955])

For any theory $\mathcal{T}$ with class of models $\mathcal{M}$, we denote by $\mathcal{M}^-$ the complement of $\mathcal{M}$ with respect to the class of all multirelations of the same arity; the theory $\mathcal{T}^-$ associated with $\mathcal{M}^-$ is known as the *complement-theory* of $\mathcal{T}$.

If $\mathcal{T}$ is finitely-axiomatizable, therefore generated by a bound formula P, then $\mathcal{M}$ is the class of models of P. Hence $\mathcal{M}^-$ is the class of models of $\neg$P and the complement-theory $\mathcal{T}^-$ is the theory generated by $\neg$P. In this case, therefore, $\mathcal{T}^{--}=\mathcal{T}$.

5.3.1. $\mathcal{T}^{--}$ *contains* $\mathcal{T}$.

$\triangleright$ Let $\mathcal{M}$ be the class of models of $\mathcal{T}$ and $\mathcal{M}^-$ its complement. $\overline{\mathcal{M}^-}$ is the intersection of all logical classes containing $\mathcal{M}^-$; hence its complement $(\overline{\mathcal{M}^-})^-$ is contained in $\mathcal{M}$. Now the class of models of $\mathcal{T}^-$ is $\overline{\mathcal{M}^-}$, and so $\mathcal{T}^{--}$ is the theory associated with $(\overline{\mathcal{M}^-})^-$, which is contained in $\mathcal{M}$. $\triangleleft$

In the general case, the class of models of $\mathcal{T}^-$ is larger than $\mathcal{M}^-$. For example, let $\mathcal{T}$ be the theory of infinite sets. Then $\mathcal{M}^-$ is the class of finite sets; but we know that the theory $\mathcal{T}^-$ associated with $\mathcal{M}^-$ is the set of all theses, and the class $\overline{\mathcal{M}^-}$ is the class of all sets. Thus $\mathcal{T}^{--}$ is the inconsistent theory, which of course properly contains $\mathcal{T}$.

It may happen that neither $\mathcal{T}$ nor $\mathcal{T}^-$ is finitely-axiomatizable. For example, let $\mathcal{T}$ be the theory associated with the sets of even finite cardinality; the class of models of $\mathcal{T}$ consists of all such sets and the infinite sets. The class of models of $\mathcal{T}^-$ consists of the sets of odd finite cardinality and the infinite sets. Neither of these classes is a logical class. It may also happen that $\mathcal{T}=\mathcal{T}^{--}$ but $\mathcal{T}$ is not finitely-axiomatizable: an example is the theory $\mathcal{T}$ of even finite cardinals and infinite cardinals.

5.3.2. *The complement-theory* $\mathcal{T}^-$ *is the strongest theory whose intersection with* $\mathcal{T}$ *is the set of all bound theses.*

$\triangleright$ Each multirelation of the given arity is either in $\mathcal{M}$ or in the complement $\mathcal{M}^-$, whose associated theories are $\mathcal{T}$ and $\mathcal{T}^-$, respectively. Thus a formula common to $\mathcal{T}$ and $\mathcal{T}^-$ is satisfied by every multirelation, and so $\mathcal{T}\cap\mathcal{T}^-$ is the set of theses.

Conversely, let $\mathcal{U}$ be a theory such that $\mathcal{F}\cap\mathcal{U}$ is the set of theses. Then no multirelation M in $\mathcal{M}^-$ can be a model of $\mathcal{T}$; thus there exists a formula P$\in\mathcal{T}$ such that P(M)$=-$. If M were not a model of $\mathcal{U}$, there would be a formula Q$\in\mathcal{U}$ such that Q(M)$=-$, so that (P$\bigvee$Q) (M)$=-$. But P$\bigvee$Q$\in\mathcal{T}\cap\mathcal{U}$, so that P$\bigvee$Q is a thesis and cannot take the value $-$.

Thus M is a model of $\mathscr{U}$. Consequently, any model of $\mathscr{T}^-$ is a model of $\mathscr{U}$, and so $\mathscr{T}^-$ is stronger than $\mathscr{U}$. ◁

5.3.3. *For any theory* $\mathscr{T}$, $\mathscr{T}^{---}=\mathscr{T}^-$.

▷ By 5.3.1, the theory $\mathscr{T}^{---}$ contains $\mathscr{T}^-$. Similarly, $\mathscr{T}$ is contained in $\mathscr{T}^{--}$, and so $\mathscr{T}\cap\mathscr{T}^{---}$ is contained in $\mathscr{T}^{--}\cap\mathscr{T}^{---}$. But by 5.3.2, both these intersections coincide with the set of bound theses. Again by 5.3.2, $\mathscr{T}^{---}$ is contained in $\mathscr{T}^-$. ◁

5.3.4. $\mathscr{T}^-$ *is the set of all formulas* P *such that* $\neg Q \vdash P$ *for every* $Q \in \mathscr{T}$.

▷ Let $P\in\mathscr{T}^-$. Then for any $Q\in\mathscr{T}$ we have $P\vee Q\in\mathscr{T}\cap\mathscr{T}^-$. Thus $P\vee Q \dashv\vdash \neg Q\Rightarrow P$ is a thesis, so that $\neg Q\vdash P$. Conversely, if P satisfies the condition, the theory generated by P, and the theory $\mathscr{T}$, intersect in the set of theses, so that $P\in\mathscr{T}^-$. ◁

Independently of the above proof, we note that the set of formulas P such that $\neg Q\vdash P$ for every $Q\in\mathscr{T}$ is closed under deduction and conjunction, so that it is indeed a theory.

5.3.5. *If there exists a formula such that* $P\in\mathscr{T}$ *and* $\neg P\in\mathscr{T}^-$, *then* $\mathscr{T}$ *is the theory generated by* P *and* $\mathscr{T}^-$ *the theory generated by* $\neg P$.

▷ Since $\neg P\in\mathscr{T}^-$, it follows that $\neg Q\vdash\neg P$ and $P\vdash Q$ for any $Q\in\mathscr{T}$. Thus the formulas in $\mathscr{T}$ are precisely the bound formulas deducible from P. ◁

It follows that *the union of* $\mathscr{T}$ *and its complement* $\mathscr{T}^-$ *is inconsistent if and only if* $\mathscr{T}$ *is finitely-axiomatizable.*

Problems. (1) If $\mathscr{T}$ is not finitely-axiomatizable, is $\mathscr{T}\cup\mathscr{T}^-$ a saturated theory?

(2) If neither $\mathscr{T}$ nor $\mathscr{T}^-$ is finitely-axiomatizable, is it true that $\mathscr{T}^{--}=\mathscr{T}$?

5.4. Categoricity

Let α be a cardinal. A theory $\mathscr{T}$ is said to be *α-categorical* if any two of its models of cardinal α are isomorphic.

5.4.1. *Let* $\mathscr{T}$ *be a theory having only infinite models and α-categorical for some cardinal α. Then* $\mathscr{T}$ *is saturated.*

▷ Let R and R′ be any two models of $\mathscr{T}$. By the denumerable-model theorem and the extension theorem 4.4.5, there exist multirelations R_α and R'_α, both of cardinal α, which are logically equivalent to R and R′, respectively. Then R_α and R'_α are both models of $\mathcal{T}$, and since they are of cardinal α they are isomorphic; it follows that R and R′ are logically equivalent. ◁

Examples. The saturated theory of the chain of rationals is $\aleph_0$-categorical, but not α-categorical for any $\alpha > \aleph_0$. Indeed, any dense denumerable chain with neither minimal nor maximal element is isomorphic to the chain of rationals, but there exist nonisomorphic chains of cardinality $\alpha > \aleph_0$ having these properties.

In fact, a more general statement is true: for any homogeneous multirelation R (see 1.7), the saturated theory of R is $\aleph_0$-categorical. Moreover, there exists a set, equipollent to the continuum, of nonisomorphic denumerable homogeneous relations ([HENS, 1972]; see Volume 1, Chapter 4, Exercise 5). Hence there exist continuum many $\aleph_0$-categorical theories (for a previous proof, see [GLA, 1970]).

Recall that an ordering is a *lattice* if any two elements have a supremum (least upper bound) and an infimum (greatest lower bound). The lattice is said to be *distributive* if the supremum and infimum operations are distributive over each other; it is *Boolean* if it is distributive, has a minimal element 0 and a maximal element 1, and moreover each element a has a complement a' such that $\inf(a, a') = 0$ and $\sup(a, a') = 1$. An *atom* is an element a such that $0 < a$ and there is no element a_1 with $0 < a_1 < a$. The saturated theory of a denumerable atomless Boolean lattice is $\aleph_0$-categorical (example: the set of finite unions of intervals of irrational numbers with rational endpoints in (0, 1)). However, the theory is not α-categorical for any $\alpha > \aleph_0$ (see Volume 1, 7.4.2).

A *field* is a birelation (sum, product) satisfying the usual conditions. The saturated theory of the field of complex numbers is α-categorical for every nondenumerable cardinal α, but not $\aleph_0$-categorical [TAR, 1951]. Indeed, a well-known theorem of Steinitz states that any two algebraically closed fields of characteristic zero and of the same infinite nondenumerable cardinal are isomorphic. This is false for denumerable fields.

The saturated theory of an algebraically closed field of prime characteristic p is α-categorical for any nondenumerable α.

The theory of a commutative group in which any sum of p equal terms (p a prime) vanishes is always categorical.

5.4.2. It can be proved that *if a theory is α-categorical for at least one cardinal $\alpha > \aleph_0$, then it is α-categorical for all such α* ([MOR, 1962]; the problem was originally posed by [ŁOS, 1954]; see also [EHR, 1957]).

We thus infer the following four possibilities for a theory whose models are infinite: (1) $\aleph_0$-categorical; (2) α-categorical for infinite nondenumerable α; (3) α-categorical for any infinite α; (4) not categorical for any cardinal. The last case is consistent with saturation, since it holds for the theory of discrete chains, say with neither minimal nor maximal element.

5.4.3. Let R and R′ be two models of a theory $\mathscr{T}$, with bases E and E′, respectively; let $(a_1, \ldots, a_n)$ and $(a'_1, \ldots, a'_n)$ be n-tuples of elements of E and E′, respectively. Let us say that these n-tuples are *logically equivalent* if the mapping taking each a_i onto a'_i $(i = 1, \ldots, n)$ is a (k, p)-isomorphism of R onto R′ for all k and p. This is readily seen to be an equivalence relation. In general, even for $n = 1$, there are infinitely many equivalence classes. For example, if R is the chain of natural numbers each class contains a single number, for each natural number a is preserved by the unique $(1, a+1)$-automorphism of R defined on $\{a\}$.

A local isomorphism satisfying the above conditions will be called a *logical isomorphism* of R onto R′; in other words, a logical isomorphism is a (k, p)-isomorphism for all k, p. A mapping which is a (k, p)-automorphism of R for all k, p will be called a *logical automorphism* (see 1.3.3).

The following theorem is due to Ehrenfeucht (see [VAU, 1961] and [SHO, 1967]). Let R be a multirelation, F a finite subset of $|R|$, such that for arbitrarily large i there exists an (i, i)-automorphism of R with domain F which is not a logical automorphism. *Then there exists a denumerable multirelation* S, *logically equivalent to* R, *such that no local isomorphism of* R *onto* S *with domain* F *is a logical isomorphism* (see 2.3.2).

With the help of this theorem we can prove the following theorem of [RYL, 1959] (see also [ENG, 1959] and [SVE, 1959]):

$\aleph_0$-CATEGORICITY THEOREM. *Let $\mathscr{T}$ be a saturated theory having only infinite models. Then the following four conditions are equivalent:*

(1) *$\mathscr{T}$ is $\aleph_0$-categorical.*

(2) *For any positive integer n, there exist only finitely many classes of logically equivalent n-tuples (for all the models of $\mathscr{T}$).*

(3) *For any model* R *of $\mathscr{T}$ and any finite subset* F *of* $|R|$, *there exist two*

natural numbers k, p such that any (k, p)-automorphism of R *with domain* F *is a logical automorphism.*

(4) *There exist a model* R *of* $\mathcal{T}$ *and, for any n, two natural numbers* k_n, p_n, *such that any* (k_n, p_n)*-automorphism of* R *defined on n elements is a logical automorphism.*

▷ (1)⇒(3). Suppose that condition (3) does not hold. Let R be a model of $\mathcal{T}$ and F a finite subset of |R| such that for arbitrarily large i there exists an (i, i)-automorphism f_i of R, with domain F, which is not a logical automorphism. We may assume without loss of generality that |R| is denumerable, replacing R if necessary by a logical restriction with a denumerable base containing $f_i(\mathrm{F})$ for each i. By Ehrenfeucht's theorem, there exists a denumerable multirelation S, which is logically equivalent to R and therefore a model of $\mathcal{T}$, such that no local isomorphism of R onto S with domain F is a logical isomorphism; thus S is not isomorphic to R and (1) is false.

(3)⇒(1). Let R and R′ be two denumerable models of $\mathcal{T}$, which are of course logically equivalent. For any finite subset F of |R|, any logical isomorphism f of R onto R′ with domain F, and any element $a \in |\mathrm{R}|$, there exists $a' \in |\mathrm{R}'|$ such that the extension of f obtained by mapping a onto a' is still a logical isomorphism; and the same holds with R and R′ interchanged. Thus, since R and R′ are denumerable, we can start from the empty mapping, which is a logical isomorphism, and successively adjoin elements, alternating between the first element of |R| not in the domain and the first element of |R′| not in the range, finally obtaining a sequence of local isomorphisms whose limit is an isomorphism of R onto R′.

(3)⇒(4). Let R be a denumerable model of $\mathcal{T}$ which satisfies (3) but not (4): there exists n such that for arbitrarily large i there is an (i, i)-automorphism of R defined on n elements which is not a logical automorphism. For each of these infinitely many indices i we let R_i and R'_i be multirelations isomorphic to R, with bases containing a common subset F of cardinality n, such that the identity mapping on F is an (i, i)-isomorphism of R_i onto R'_i but not a logical isomorphism. Let $a_1, \ldots, a_n$ denote the elements of F; for each i, let $\mathrm{A}^i_1, \ldots, \mathrm{A}^i_n$ and $\mathrm{A}'^i_1, \ldots, \mathrm{A}'^i_n$ denote the corresponding singleton relations with bases $|\mathrm{R}_i|$ and $|\mathrm{R}'_i|$, respectively. By confining attention to a suitable subsequence of the i's, we may assume that the multirelations $\mathrm{R}_i\mathrm{A}^i_1 \ldots \mathrm{A}^i_n$ converge logically to a denumerable multirelation $SB_1 \ldots B_n$ and the multirelations $R'_iA'^i_1 \ldots A'^i_n$ to a denumerable multirelation $S'B'_1 \ldots B'_n$ (apply 2.2.2). We may assume moreover that the bases of the limiting multirelations contain F and that the relations B_i and B'_i are the singletons of the corresponding elements a_i. Then, for each j, the identity mapping on F becomes a (j, j)-isomorphism of $R_iA^i_1 \ldots A^i_n$ onto $SB_1 \ldots B_n$ for all sufficiently large i; a similar statement holds with R, A, S, B replaced by R′, A′, S′, B′. It follows that the identity mapping on F is a logical isomorphism of S onto S′. Now we have already proved that (3)⇔(1), and so S and S′ are both isomorphic to R; hence, by (3), for sufficiently large j any (j, j)-isomorphism from S onto an isomorphic image, defined on F, is a logical isomorphism. In particular, for sufficiently large i the identity mapping on F is a logical isomorphism of S onto R_i and of S′ onto R'_i, thus also of S onto S′ and, finally, of R_i onto R'_i – contradiction.

(4)⇒(3). Let R be a model satisfying (4). It will suffice to show that any other model R′ of $\mathscr{T}$ (which is of course logically equivalent to R) satisfies (4) with the same numbers k_n, p_n for each n. Indeed, let f be a (k_n, p_n)-automorphism of R′ defined on n elements. For each $i \geqslant k_n$, $i \geqslant p_n$, we apply an (i, i)-isomorphism of R′ onto R to the domain and range of f; this shows that f is also an (i, i)-automorphism of R′, and hence, finally, a logical automorphism of R′.

(4)⇒(2). Consider all systems $(R; a_1, \ldots, a_n)$, where $a_1, \ldots, a_n \in |R|$, and partition them into equivalence classes, putting $(R; a_1, \ldots, a_n)$ and $(R'; a'_1, \ldots, a'_n)$ into the same class if the mapping taking each a_i onto a'_i $(i = 1, \ldots, n)$ is a logical isomorphism of R onto R′. By (4), any one class is fully determined by its (k_n, p_n)-equivalence class, and by 1.2.1 there are only finitely many such classes. Thus the systems in question fall into a finite number of logical equivalence classes.

(2)⇒(4). Let R be a model of $\mathscr{T}$ and suppose that there are only finitely many logical equivalence classes of n-tuples. For any two n-tuples from distinct classes there exist k and p such that the mapping of one n-tuple onto the other is not a (k, p)-automorphism of R. Considering these values of k and p for all pairs of classes, we let k_n and p_n be their greatest values. Condition (4) is then satisfied. ◁

5.4.4. In connection with categoriticity, it is worth noting that *no logical equivalence class can contain two nonisomorphic denumerable multirelations*

[VAU, 1961 and 1963]. As an example pertaining to the cardinal of the continuum, consider the class of binary equivalence relations with two infinite equivalence classes. Here either one class is denumerable and the other equipollent to the continuum, or both are equipollent to the continuum.

There exists a logical equivalence class comprising exactly three denumerable relations. The following example was first given by Ehrenfeucht, using infinite sequences of relations, and subsequently, in the form presented here, by [VAU, 1961]. We construct a relation R as follows. Let Q_i be the relation obtained from the chain Q of rationals by replacing each element by i incomparable elements, $i = 1, 2, \ldots$. Then define R as the sum $Q_1 + Q_2 + \cdots$. Let R′ be the extension of R consisting of R followed by Q_∞ (the chain Q with each element replaced by a denumerably infinite set of pairwise incomparable elements). Finally, construct an extension R″ of R′ by adding ("after" R) the relation obtained by replacing each element of $1 + Q$ by a denumerably infinite set of pairwise incomparable elements. Then R′ is a logical extension of R and R″ a logical extension of R′. It can be shown that these are the only relations with denumerable base which are logically equivalent to R (up to isomorphism). Analogous examples may be adduced for any natural number $\geqslant 3$.

5.5. Model-saturated theory

A theory $\mathscr{T}$ is said to be *model-saturated* if, for any two models R, R′ of $\mathscr{T}$ such that R′ is an extension of R, R′ is a logical extension of R (see [A. ROB, 1965], p. 91).

Examples. The theory of dense chains with neither minimal nor maximal element is saturated and model-saturated. On the other hand, the theory of dense chains with minimal element but no maximal element is saturated but not model-saturated: take the chain of rationals $\geqslant -1$ as an extension of the chain of rationals $\geqslant 0$.

The theory of dense positive sums (sum relation over nonnegative rational numbers), the theory of discrete divisible positive sums (natural numbers), and the theory of real fields are both saturated and model-saturated (see 3.4, 3.5, 3.6).

A theory may be model-saturated but not saturated. An example is the theory of algebraically closed fields. For example, the formula stating that the characteristic of the field is 2 ($a+a=0$ for every a) is not in the theory, and neither is its negation, so that the theory is not saturated; it is, however, model-saturated (see [A. ROB, 1965], p. 102).

5.5.1. MODEL-SATURATION TEST ([A. ROB, 1965], p. 92). *$\mathcal{T}$ is model-saturated if and only if, for any two models* R, R′ *of $\mathcal{T}$ such that* R′ *is an extension of* R, R′ *is a* 1*-extension of* R. *In other words, for any finite subset* F *of* |R′| *there exists an isomorphism of* R′ | F *onto a restriction of* R *which is the identity on* F∩|R| (see 4.3.5).

▷ Every logical extension is *a fortiori* a 1-extension; hence, if $\mathcal{T}$ is model-saturated it satisfies the condition.

Conversely, suppose that $\mathcal{T}$ is not model-saturated; in other words, it has two models R, R′ such that R′ is an extension of R but not a logical extension. Then there exist k, p and a finite subset F of |R| such that the identity mapping on F is not a (k, p)-isomorphism of R onto R′. We may assume that the models R, R′ have been so chosen that k is minimal; note that $k\geqslant 1$, since R | F = R′ | F.

For any finite superset F̄ of F contained in |R|, the identity mapping on F̄ is a $(k-1, r)$-isomorphism of R onto R′ for all r, since k is minimal. Since the identity mapping on F is not a (k, p)-isomorphism, there exist a number $q\leqslant p$ and a finite set G obtained by adding q elements of |R′| to F such that no bijective mapping defined on G whose restriction to F is the identity is a $(k-1, p-q)$-isomorphism of R′ onto R.

If $\mathcal{T}$ satisfies the condition, R′ is a 1-extension of R. Thus, for any finite superset H of G contained in |R′| there exists a local isomorphism of R′ onto R whose restriction to F is the identity mapping, so that is not a $(k-1, p-q)$-isomorphism. In other words, up to isomorphism R is an extension of R′ | H and a model of $\mathcal{T}$, and the identity mapping on G is not a $(k-1, p-q)$-isomorphism of R′ onto this extension. By the sharpened version of the Henkin extension theorem 4.4.4.2, there exists an extension R″ of R′ which is a model of $\mathcal{T}$, such that the identity mapping on G is not a $(k-1, p-q+h)$-isomorphism of R′ onto R″ (where $h=\text{card G}$); this contradicts the minimality of k, and so the theory $\mathcal{T}$ cannot satisfy the condition. ◁

5.5.2. *Let $\mathcal{T}$ be a model-saturated theory,* R *a model of $\mathcal{T}$, and* F *a finite* *subset of the base* $|R|$. *For arbitrary natural numbers* k, p, *there exists a natural number* q *with the following property: given any model* R' *of* $\mathcal{T}$ *and any local isomorphism* f *of* R *onto* R' *defined on* F, *if* f *may be extended to any* q *elements of* $|R|-F$, *then* f *is a* (k, p)*-isomorphism of* R *onto* R'.

▷ Suppose the contrary. Let k, p be such that for any q there exist a model R' of $\mathcal{T}$ and a local isomorphism f of R onto R' which is extendible to any q elements of $|R|-F$ but is not a (k, p)-isomorphism. Let G be a finite superset of F in $|R|$. Letting q be the cardinality of $G-F$, we see that there exists a model of $\mathcal{T}$ which is an extension of $R \mid G$, say S, such that the identity mapping on F is not a (k, p)-isomorphism of R onto S. By 4.4.4.2, there exists an extension R' of R, which is a model of $\mathcal{T}$, such that the identity mapping on F is not a $(k, p+h)$-isomorphism of R onto R' (where $h=\text{card}\, F$). But $\mathcal{T}$ is model-saturated; thus R' is a logical extension of R and so the identity mapping on F is a logical isomorphism of R onto R' – contradiction. ◁

5.5.3. THEOREM (ELIMINATION OF QUANTIFIERS IN MODEL-SATURATED THEORIES).

(1) *Let* $\mathcal{T}$ *be a model-saturated theory,* P *an* n*-ary logical formula which is* $+$ *for at least one model* R *of* $\mathcal{T}$ *and a sequence of* n *elements* $a_1, \ldots, a_n$ *of* $|R|$. *Then there exists an* n*-ary existential prenex formula* Q *whose truth value for any model* R' *of* $\mathcal{T}$ *and any sequence* $a'_1, \ldots, a'_n$ *of elements of* $|R'|$ *is*

$$Q(R')(a'_1, \ldots, a'_n) = P(R')(a'_1, \ldots, a'_n).$$

(2) *Conversely, any theory* $\mathcal{T}$ *satisfying the above condition for every formula* P *is model-saturated.*

▷ (1) Let (k, p) be the characteristic of P. Consider all systems $M=(R; a_1, \ldots, a_n)$, where R is a model of $\mathcal{T}$ such that $P(R)(a_1, \ldots, a_n)=+$. Partition these systems into (k, p)-equivalence classes. Take a representative $M=(R; a_1, \ldots, a_n)$ of each class, and let q_M be the natural number associated with M by 5.5.2. Let Q_M be an existential prenex formula stating the existence (up to isomorphism) of all extensions of $R \mid \{a_1, \ldots, a_n\}$ to q_M arbitrary new elements of $|R|$. By 5.5.2, for any model R' of $\mathcal{T}$ and any $a'_1, \ldots, a'_n$ in $|R'|$, the equality $Q_M(R')(a'_1, \ldots, a'_n)=+$ implies that $P(R')(a'_1, \ldots, a'_n)=+$. Letting the (k, p)-equivalence classes vary (there are only finitely many), and considering all systems $M=(R; a_1, \ldots, a_n)$ in each class, we obtain the n-ary logical class represented by P as a union of infinitely many classes (those defined by the formulas Q_M). By the compactness theorem, the logical class represented by P is the union of a finite number of these classes. The disjunction of all the corresponding formulas Q_M is the desired existential prenex formula.

(2) Let $\mathcal{T}$ be a theory satisfying the condition of part 1 for every formula P. Let R, R′ be two models of $\mathcal{T}$, such that R′ is an extension of R. Let P be an n-ary formula and $a_1, \ldots, a_n$ elements of $|R|$ such that $P(R)\ (a_1, \ldots, a_n) = +$. By assumption, there exists an n-ary existential prenex formula Q such that $Q(R)\ (a_1, \ldots, a_n) = +$; hence also $Q(R')\ (a_1, \ldots, a_n) = +$, since R′ is an extension of R. Moreover, since Q satisfies the conditions of (1), we have $P(R)\ (a_1, \ldots, a_n) = +$. Thus the identity mapping on a finite subset of $|R|$ preserves the truth value of any formula, and so R′ is a logical extension of R. ◁

5.5.4. *Let* R, R′ *be two models of a model-saturated theory* $\mathcal{T}$. *If every finite restriction of* R *is embeddable in* R′, *then* R *and* R′ *are logically equivalent.*

▷ There exists an extension S of R which is logically equivalent to R′ (see 4.4.3). S is a model of $\mathcal{T}$, therefore a logical extension of R. Thus S is logically equivalent to both R and R′. ◁

5.5.5. Any model-saturated theory is closed under passage to limits; this follows immediately from the definition (5.5) and from theorem 1.4.1.3. Conversely, a theory which is closed under passage to limits and α-categorical for at least one infinite cardinal α, and has only infinite models, is necessarily model-saturated (see [LIN, 1964]).

Hint. Call a multirelation R 1-*extensive* if any extension of R with the same age (i.e., the same finite restrictions, up to isomorphism) is a 1-extension of R. Then any R has a 1-extensive extension of the same cardinal. If $\mathcal{T}$ is an α-categorical theory, any model of $\mathcal{T}$ of cardinal α is 1-extensive. On the other hand, any logical restriction of a 1-extensive multirelation is 1-extensive. Thus all denumerable models of $\mathcal{T}$ are 1-extensive. By 5.5.1, with the additional assumption that R and R′ are denumerable, $\mathcal{T}$ is model-saturated.

EXERCISES

1

(1) Let $\mathscr{T}$, $\mathscr{U}$ be two theories and $\mathscr{M}$, $\mathscr{N}$ their classes of models. Let $\mathscr{M} \subseteq \mathscr{N}$, so that $\mathscr{T} \supseteq \mathscr{U}$; let $\mathscr{M}'$ be the complement of $\mathscr{M}$. Then the theory associated with the intersection $\mathscr{M}' \cap \mathscr{N}$ is known as the complement of $\mathscr{T}$ modulo $\mathscr{U}$, denoted by $\mathscr{T}_{\mathscr{U}}$. Show that $\mathscr{T} \cap \mathscr{T}_{\mathscr{U}} = \mathscr{U}$; in fact, $\mathscr{T}_{\mathscr{U}}$ is the strongest theory whose intersection with $\mathscr{T}$ is $\mathscr{U}$. Note that if $\mathscr{U}$ is the set of theses the complement modulo $\mathscr{U}$ is simply the complement of a theory as defined in 5.3.

(2) Given any two theories $\mathscr{T}$, $\mathscr{U}$, let us write $\mathscr{T} \vdash \mathscr{U}$ instead of $\mathscr{T} \subseteq \mathscr{U}$. Call $\mathscr{T}$ a "thesis" if it is the set of all formulas (i.e., is inconsistent), and an "antithesis" if it is the set of theses. Let $\neg\mathscr{T}$ denote the complement-theory of $\mathscr{T}$, $\mathscr{T} \bigwedge \mathscr{U}$ the intersection $\mathscr{T} \cap \mathscr{U}$, and $\mathscr{T} \bigvee \mathscr{U}$ the union-theory of $\mathscr{T}$ and $\mathscr{U}$. Finally, $\mathscr{T} \Rightarrow \mathscr{U}$ will denote the theory $\mathscr{T}_{\mathscr{T} \cap \mathscr{U}}$ (the complement of $\mathscr{T}$ modulo $\mathscr{T} \cap \mathscr{U}$). The "connective calculus" thus constructed, with four connections $\neg$, $\bigwedge$, $\bigvee$, $\Rightarrow$, is the *intuitionistic calculus* of [HEY, 1930].

Show that $\mathscr{T} \bigwedge \neg\mathscr{T}$ is an "antithesis", $\mathscr{T} \Rightarrow \mathscr{T}$ is a "thesis". However, $\mathscr{T} \bigvee \neg\mathscr{T}$ need not be a "thesis": in other words, the principle of the excluded middle fails to hold. It is always true that $\mathscr{T} \vdash \neg\neg\mathscr{T}$, but not that $\neg\neg\mathscr{T} \vdash \mathscr{T}$. Another true statement is $\mathscr{T} \bigwedge (\mathscr{T} \Rightarrow \mathscr{U}) \vdash \mathscr{U}$. The "theses" in $\neg$, $\bigwedge$ are the same as in the classical connective calculus.

2

Let $\mathscr{T}$ be a theory and P a bound formula of the same predicarity. We shall say that P is *model-consistent* with $\mathscr{T}$ if, for any model R of $\mathscr{T}$, there exists an extension of R which is a model of both $\mathscr{T}$ and P (see [A. ROB, 1965], p. 84).

(1) Note that if P is model-consistent with $\mathscr{T}$, then P is consistent with $\mathscr{T}$ in the sense that $\mathscr{T} \cup \{P\}$ generates a consistent theory. On the other hand, let $\mathscr{T}$ be the theory of chains (total orderings) and P the formula $\forall_{xy}\, x \equiv y$; then P is consistent with $\mathscr{T}$ but not model-consistent. The formula stating that the chain is dense (or discrete) is model-consistent with $\mathscr{T}$.

(2) Show that for a given theory $\mathscr{T}$ consistency coincides with model-consistency (for any formula added to $\mathscr{T}$) if and only if, for any model R of $\mathscr{T}$ and any formula P, either P is in $\mathscr{T}$, or $\neg$P is in $\mathscr{T}$, or R has an extension which is a model of $\mathscr{T}$ and P and another extension which is a model of $\mathscr{T}$ and $\neg$P. This is the case for any saturated theory; another example is the theory of dense chains (no stipulation being made as to the existence of minimal and/or maximal elements). A theory satisfying the above condition has either exactly one model with finite base (up to isomorphism) or only models with infinite bases.

(3) Show that for a given theory $\mathscr{T}$ consistency coincides with model-consistency if and only if $\mathscr{T}$ is *universally saturated*, in the following sense: for any universal bound formula (prenex formula containing only universal quantifiers), either P or $\neg$P is in $\mathscr{T}$ (this result is due to R. Cusin, 1971). Hence deduce that for the theory of infinite chains consistency and model-consistency coincide.

(4) Call a theory $\mathscr{T}$ *weakly universally saturated* [RIB, 1961] if, for any universal bound formula P, either P is in $\mathscr{T}$ or there exists a natural number p such that $\neg P \Leftrightarrow A_p$ (see 5.1.2). Examples: the theory of chains, the theory of Boolean lattices. Show that the following is an equivalent definition: for any finite restriction R of a model of $\mathscr{T}$, there exists p such that R is embeddable in any model of $\mathscr{T}$ of cardinality $\geqslant p$.

3

Given any theory $\mathscr{T}$, let $\mathscr{T}^{U}$ be the theory generated by all universal formulas in $\mathscr{T}$. Prove that a multirelation R is a model of $\mathscr{T}^{U}$ if and only if R has an extension which is a model of $\mathscr{T}$.

Hint: It is obvious that if R is a restriction of a model of $\mathscr{T}$, then R is a model of $\mathscr{T}^{U}$. Conversely, let R be a model of $\mathscr{T}^{U}$. By 4.4.4.1, it suffices to show that for any finite subset $F \subseteq |R|$ there is an extension of $R \mid F$ which is a model of $\mathscr{T}$. Supposing the contrary, let p be the cardinality of F. Construct a free formula A such that $\underset{1,\dots,p}{\exists}$ A takes the value + for extensions of $R \mid F$ (and only for these, up to isomorphism). Then $\mathscr{T}$ and $\underset{1,\dots,p}{\exists}$ A are incompatible. By the compactness theorem, there exists a formula P in $\mathscr{T}$ such that $P \vdash \underset{1,\dots,p}{\forall} \neg A$. The latter formula is in $\mathscr{T}^{U}$, and so R is not a model of $\mathscr{T}^{U}$.

4

Call $\mathscr{T}$ a *filtering theory* if any two models of $\mathscr{T}$ have a common extension which is again a model of $\mathscr{T}$. Prove that $\mathscr{T}$ is a filtering theory if and only if, for any two existential formulas P, P′, if $\mathscr{T} \cup \{P\}$ and $\mathscr{T} \cup \{P'\}$ are consistent, then so is $\mathscr{T} \cup \{P, P'\}$.

Hint: Necessity is obvious. To prove sufficiency, suppose that $\mathscr{T}$ is not a filtering theory and let R, R′ be two models having no common extension which is a model of $\mathscr{T}$. By 4.4.8.2, we can find finite sets $F \subseteq |R|$ and $F' \subseteq |R'|$ such that $R \mid F$ and $R' \mid F'$ have no common extension which is a model of $\mathscr{T}$. Let $p = \text{card}\, F$, $p' = \text{card}\, F'$; let A, A′ be free formulas such that $P = \underset{1,\dots,p}{\exists}$ A is + for extensions of $R \mid F$ and $P' = \underset{1,\dots,p}{\exists}$ A′ is + for extensions of $R' \mid F'$ (up to isomorphism). Then $\mathscr{T} \cup \{P, P'\}$ is inconsistent.

5

Show that the saturated theory of the sum of rational numbers is α-categorical for any nondenumerable cardinal α, but not $\aleph_0$-categorical. The same is true for the saturated theory of the succession relation on the natural numbers.

6

Using Exercise 3, show that two model-saturated theories with the same universal formulas are identical. For any theory $\mathscr{T}$, the only possible model-saturated theory with the same universal formulas as $\mathscr{T}$ is known as the *model-companion* of $\mathscr{T}$.

CHAPTER 6

PSEUDO-LOGICAL CLASS; INTERPRETABILITY OF THEORIES; EXPANSION OF A THEORY; AXIOMATIZABILITY

6.1. Pseudo-logical class

Let ρ, σ be two arities (finite sequences of natural numbers). Let $\mathscr{A}$ be a logical class of arity (ρ, σ), i.e., a class of concatenations (R, S) where R is a ρ-ary multirelation and S a σ-ary multirelation. The class of multirelations R for which there exists S such that (R, S) is in $\mathscr{A}$ is called an $\mathscr{P}$-*logical class* relative to $\mathscr{A}$. The class of all R such that (R, S) is in $\mathscr{A}$ for *any* S with the same base is known as an $\mathscr{P}^{\prime}$-*logical class* relative to $\mathscr{A}$. Both types are known as *pseudo-logical classes.*

Any logical class is pseudo-logical (let σ be the empty sequence). We now give an example of a pseudo-logical class which is not a logical class. Let I be the chain of natural numbers and C the succession relation on the natural numbers. The class $\mathscr{A}$ of birelations which are logically equivalent to (C, I) is a logical class (see 1.5.2). The class $\mathscr{C}$ of relations C′ for which there exists a relation I′ such that (C′, I′) is in $\mathscr{A}$ is a pseudo-logical class, but not a logical class. Indeed, by 1.5.3, for any k and p there exists a succession relation C″ which contains a finite cycle and is (k, p)-equivalent to C, but is not in $\mathscr{C}$.

A pseudo-logical class need not even be closed with respect to logical equivalence. For example, let I be the chain of natural numbers and S the sum of natural numbers; let $\mathscr{A}$ be the class of birelations (I′, S′) which are (2, 6)-equivalent to (I, S). Let $\mathscr{I}$ be the pseudo-logical class of relations I′ for which there exists S′ such that (I′, S′) is in $\mathscr{A}$. Then the chain I″ consisting of the natural numbers followed by the integers is logically equivalent to I (see 1.4). However, I″ is not in $\mathscr{I}$: there is no relation S″ such that (I″, S″) is in $\mathscr{A}$ (see 1.4.2).

Another example (in this case an $\mathscr{P}^{\prime}$-logical class) is the class of all ordinals (and their isomorphic images): R is an ordinal if and only if it is a chain and, for any unary relation S with the same base, the set of elements on which S takes the value + has an R-minimal element. Here again the class is not even closed under logical equivalence: the ordinal ω is logically equivalent to $\omega+\zeta$, where ζ is the chain of integers.

6.1.1. *The union and intersection of two $\mathscr{I}$-logical classes is an $\mathscr{I}$-logical class. The same is true of $\mathscr{F}$-logical classes.*

▷ Let $\mathscr{A}$ be a logical class, and consider all multirelations R for which there exists S such that $\mathrm{RS}\in\mathscr{A}$. Similarly, let $\mathscr{B}$ be some other logical class and consider all R for which there exists T such that $\mathrm{RT}\in\mathscr{B}$. The class $\mathscr{A}^*$ of all multirelations RST, where $\mathrm{RS}\in\mathscr{A}$ and T is arbitrary, is a logical class; the same holds for the class $\mathscr{B}^*$ of all multirelations RST, where $\mathrm{RT}\in\mathscr{B}$ and S is arbitrary. Now the required union is the class of all multirelations R for which there exists ST such that $\mathrm{RST}\in\mathscr{A}^*\cup\mathscr{B}^*$. Similar reasoning yields the assertion for the intersection and for $\mathscr{F}$-logical classes. ◁

The complement of an $\mathscr{F}$-logical class is an $\mathscr{I}$-logical class, and conversely.

6.1.2. *Let $\mathscr{C}$ be an $\mathscr{I}$-logical class. For any multirelation* $\mathrm{R}\in\mathscr{C}$ *and any superset* E′ *of its base* E *such that the cardinal of* $\mathrm{E}'-\mathrm{E}$ *is at least that of* E, *there exists a logical extension* R′ *of* R to E′ *such that* $\mathrm{R}'\in\mathscr{C}$.

▷ Suppose that $\mathscr{C}$ is $\mathscr{I}$-logical relative to some logical class $\mathscr{A}$. Thus, there exists a multirelation S such that $\mathrm{RS}\in\mathscr{A}$. Using the logical extension theorem 4.4.5, consider a logical extension R′S′ of RS with base E′. Then $\mathrm{R'S'}\in\mathscr{A}$, so that $\mathrm{R}'\in\mathscr{C}$. Moreover, R′ is a logical extension of R. ◁

Consequently, *the class of ordinals* (and their isomorphic images) *is an $\mathscr{F}$-logical class but not an $\mathscr{I}$-logical* class. Indeed, every logical extension of the ordinal ω is a discrete chain obtained by adjoining to ω isomorphic images of the chain of integers; this is certainly not an ordinal.

6.1.3. *A pseudo-logical class which is both $\mathscr{F}$-logical and $\mathscr{I}$-logical is a logical class* (communicated by R. Laver).

▷ Let ρ, σ, τ be three sequences of predicates, $\mathscr{A}$ a logical class of arity $\rho\sigma$, and $\mathscr{B}$ a logical class of arity $\rho\tau$. Let $\mathscr{F}_\sigma\mathscr{A}$ be the class of multirelations R such that, for any S (assignable to σ) $\mathrm{RS}\in\mathscr{A}$, and $\mathscr{I}_\tau\mathscr{B}$ the class of all R such that $\mathrm{RT}\in\mathscr{B}$ for some T assignable to τ. Suppose that $\mathscr{F}_\sigma\mathscr{A}=\mathscr{I}_\tau\mathscr{B}$. Consider the logical class $\mathscr{A}^*$ of arity $\rho\sigma\tau$ consisting of all multirelations RST such that $\mathrm{RS}\in\mathscr{A}$ and T is arbitrary; similarly, let $\mathscr{B}^*$ be the class of all RST such that RT$\in\mathscr{B}$ and S is arbitrary. Finally, let $\mathscr{C}$ denote the class of all multirelations RST with R$\in\mathscr{P}_\sigma\mathscr{A}=\mathscr{P}_\tau\mathscr{B}$ and S, T arbitrary. It is clear that $\mathscr{B}^*\subseteq\mathscr{C}\subseteq\mathscr{A}^*$. Moreover, any subclass of $\mathscr{A}^*$ in which σ and τ are inactive must also be a subclass of $\mathscr{C}$, and any class containing $\mathscr{B}^*$ in which σ and τ are inactive must also contain $\mathscr{C}$. By the interpolation theorem (Volume 1, 6.7.2), there exists a logical class between $\mathscr{B}^*$ and $\mathscr{A}^*$ in which σ and τ are inactive. By the preceding reasoning, this class must be $\mathscr{C}$, which is therefore a logical class, as required, when viewed as a class of arity ρ. ◁

6.2. Interpretability of Theories

Let $\mathscr{T}$ be a theory, $\mathscr{M}$ its class of models, and $\mathscr{P}_1,\ldots,\mathscr{P}_h$ a finite sequence of logical operators of the same predicarity as the formulas of $\mathscr{T}$. For each model R in $\mathscr{M}$, we consider the multirelation $\mathscr{P}_1(\mathrm{R}),\ldots,\mathscr{P}_h(\mathrm{R})$. The theory associated with the class of these multirelations will be called *the theory interpretable in* $\mathscr{T}$ *by* $\mathscr{P}_1,\ldots,\mathscr{P}_h$.

An equivalent definition is as follows. Consider h logical formulas of the form

$$\underset{1\ldots m}{\forall}\ \rho x^1\ldots x^m\Leftrightarrow \mathrm{P},$$

where P is a formula of arity m; we call these formulas *definitions*. Take the theory generated by $\mathscr{T}$ and the definitions, and omit all formulas containing predicates other than the h new predicates ρ. This definition is *noncreative*, in the sense that any formula Q in which ρ is inactive which belongs to the theory generated by $\mathscr{T}$ and the definition of ρ is already in $\mathscr{T}$. In particular, if $\mathscr{T}$ is consistent, it remains consistent after addition of definitions. The definition is also *eliminable*, that is, for any formula Q containing ρ, there exists a formula Q′ in which ρ is inactive such that Q$\Leftrightarrow$Q′ is in the theory generated by $\mathscr{T}$ and the definition of ρ.

The concepts of noncreativity and eliminability were anticipated by Pascal and made rigorous by [LES, 1927]; see also [SUP, 1960].

An obvious example of interpretability is obtained by restricting a given theory $\mathscr{T}$ to formulas which contain only some of the predicates of $\mathscr{T}$, or alternatively by duplicating certain predicates (ρ is duplicated by ρ' if one adds the definition

$$\underset{1\ldots m}{\forall}\ \rho x^1\ldots x^m\Leftrightarrow\rho' x^1\ldots x^m).$$

As another example, let $\mathscr{T}$ be the theory formed by the set of free theses with no predicates; let $\mathscr{P}$ be the logical operator which associates with every base E the unary relation which is + for the elements of E. The result is the theory generated by $\underset{x}{\forall}\,\rho x$.

6.2.1. A theory $\mathscr{U}$ is interpretable in $\mathscr{T}$ by operators $\mathscr{P}$ if and only if the following two conditions hold:

(1) For each model R of $\mathscr{T}$, the transform $\mathscr{P}(\mathrm{R})$ is a model of $\mathscr{U}$.

(2) For any logical class $\mathscr{A}$, if the inverse image $\mathscr{P}^{-1}(\mathscr{A})$ is in $\mathscr{T}$ (i.e., is represented by a formula of $\mathscr{T}$), then $\mathscr{A}$ is in $\mathscr{U}$ (and conversely, by condition 1).

6.2.2. Interpretability is a pre-ordering on the class of theories (to prove this, one uses 6.2.1 and composition of logical operators; see Volume 1, 5.2.2). The theories interpretable in a given theory form a denumerably infinite set, since each of them is defined by a sequence of logical operators. Any theory interpretable in a consistent theory is consistent.

6.2.3. *Any theory interpretable in a saturated theory is saturated.* Indeed, let $\mathscr{T}$ be the saturated theory of a multirelation R. Then the class in question consists of all the multirelations which are logically equivalent to $\mathrm{S}=\mathscr{P}_1(\mathrm{R}),\ldots,\mathscr{P}_h(\mathrm{R})$; the resulting theory is the saturated theory of S. We are thus brought back to *interpretability of multirelations* in the sense of Volume 1, 7.1.

More precisely: a theory $\mathscr{U}$ is interpretable in a saturated theory $\mathscr{T}$ if and only if one of the following equivalent conditions is satisfied: (1) There exist a model R of $\mathscr{T}$ and a model S of $\mathscr{U}$ such that S is interpretable by R. (2) For any model R of $\mathscr{T}$, there exists a model S of $\mathscr{U}$ which is interpretable by R.

6.2.4. If $\mathscr{U}$ is a theory interpretable in $\mathscr{T}$, a model of $\mathscr{U}$ need not be of the form $\mathscr{P}_1(\mathrm{R})\ldots\mathscr{P}_h(\mathrm{R})$, where R is a model of $\mathscr{T}$. For example, let $\mathscr{T}$ be the saturated theory of the sum S of natural numbers and $\mathscr{U}$ the saturated theory of the chain I of natural numbers. $\mathscr{U}$ is interpretable in $\mathscr{T}$ by the operator $\mathscr{P}$ represented by $\underset{z}{\exists}\,\sigma xzy$ (where the predicate σ is replaceable by S). Now the chain I′ defined as I followed by an isomorphic image of the chain of all integers is logically equivalent to I and is therefore a model of $\mathscr{U}$. There exists no relation S′ logically equivalent to S such that $\mathscr{P}$(S′)=I′. Indeed, if such a relation existed, (I′, S′)=($\mathscr{P}$(S′), S′) would be logically equivalent to (I, S)=($\mathscr{P}$(S), S), contradicting 1.4.2. However, we have the following theorem:

If a theory $\mathscr{U}$ is interpretable in $\mathscr{T}$ by operators $\mathscr{P}$, then for any model S *of* $\mathscr{U}$ *there exists a model* R *of* $\mathscr{T}$ *such that* $\mathscr{P}$(R) *is logically equivalent to* S.

▷ By the definition of interpretability, every model S of $\mathscr{U}$ is the logical limit of a sequence of multirelations $\mathscr{P}(R_i)$ ($i=1, 2, \ldots$), where each R_i is a model of $\mathscr{T}$. A suitably chosen subsequence of $\{R_i\}$ converges to a multirelation R, which is necessarily a model of $\mathscr{T}$. Thus the sequence $\mathscr{P}(R_i)$ converges to $\mathscr{P}$(R), and also to S. ◁

6.2.5. Let $\mathscr{T}$, $\mathscr{U}$ be two theories. If every model of $\mathscr{U}$ is interpretable in a model of $\mathscr{T}$, this does not necessarily imply that $\mathscr{U}$ is interpretable in $\mathscr{T}$. Let R, S be two relations with the same finite base and the same arity, but not isomorphic. Let $\mathscr{T}$ be the saturated theory of the birelation (R, S) and $\mathscr{U}$ the intersection of the saturated theories of R and S. Since the base is finite, there exists a formula taking the value + only for relations isomorphic either to R or to S, and $\mathscr{U}$ is the set of formulas deducible from this formula; thus every model of $\mathscr{U}$ is isomorphic to R or to S. Hence any model of $\mathscr{U}$ is interpretable in (R, S) and therefore in a model of $\mathscr{T}$. But $\mathscr{U}$ is not saturated and so cannot be interpretable in the saturated theory $\mathscr{T}$.

6.2.6. If $\mathscr{T}$ and $\mathscr{U}$ are saturated theories, each interpretable in the other, there need not exist a model of $\mathscr{T}$ and a model of $\mathscr{U}$ each of which is interpretable in the other (unpublished result of A. Roberty, 1970; see Exercise 3).

Problem. If $\mathscr{T}$ and $\mathscr{U}$ are saturated theories, each interpretable in the other, such that $\mathscr{T}$ is finitely-axiomatizable, is $\mathscr{U}$ finitely-axiomatizable? The same question may be asked for nonsaturated theories.

6.3. Canonical expansion, semantic expansion, and other expansions

Let P be a formula in predicates ρ, and let σ be new predicates. We define a *canonical transform* of P, denoted by $P_{\rho,\sigma}$, to be any formula in predicates

ρ, σ, where the predicates σ are inactive, which takes the same value as P for any relations assignable to ρ.

Let $\mathscr{T}$ be a theory in predicates ρ, and σ new predicates. Let $\mathscr{T}_{\rho,\sigma}$ denote the theory in predicates ρ, σ generated by the formulas $P_{\rho,\sigma}$ for all formulas P in $\mathscr{T}$; we shall call $\mathscr{T}_{\rho,\sigma}$ the *canonical expansion* of $\mathscr{T}$ to ρ, σ.

A theory $\mathscr{T}^*$ in predicates ρ, σ is called an *expansion* of $\mathscr{T}$ if the theory generated by the formulas of $\mathscr{T}^*$ in which σ is inactive is the canonical expansion of $\mathscr{T}$. In other words, the formulas of $\mathscr{T}^*$ in which σ is inactive are precisely the canonical transforms of the formulas of $\mathscr{T}$.

Examples. If $\mathscr{T}$ consists of the bound formulas which are theses in ρ, its canonical expansion consists of all bound formulas which are theses in ρ, σ.

A theory may also have non-canonical expansions, with the same additional predicates. Let $\mathscr{T}$ be the set of theses in ρ; let σ be a unary predicate and $\mathscr{T}^*$ the set of all bound formulas in ρ, σ which are deducible from $\forall \sigma x$. Any of the latter formulas in which σ is inactive is a thesis (replace σ by a unary relation which is always +).

It follows that $\mathscr{T}^*$ is an expansion of $\mathscr{T}$, but it is distinct from the canonical expansion, which consists of the theses in ρ, σ.

Let $\mathscr{T}$ be a theory in predicates ρ. Given new predicates σ, distinct from ρ, a theory $\mathscr{T}^*$ in ρ, σ is called a *semantic expansion* of $\mathscr{T}$ if the following condition holds: any multirelation R assignable to ρ is a model of $\mathscr{T}$ if and only if there is a multirelation S assignable to σ such that RS is a model of $\mathscr{T}^*$. The above examples are all semantic expansions. Another example of a semantic expansion is the canonical expansion, whose models are all multirelations RS where R is a model of $\mathscr{T}$ and S is arbitrary.

6.3.1. *Any semantic expansion is an expansion.*

$\triangleright$ Let $\mathscr{T}$ be a theory in predicates ρ, and $\mathscr{T}^*$ a semantic expansion in predicates ρ, σ. Let P be a formula of $\mathscr{T}$; then any model R of $\mathscr{T}$ is a model of P, i.e., P(R) = +. Any canonical transform $P_{\rho,\sigma}$ is such that $P_{\rho,\sigma}(RS) = +$ for any S. Thus $P_{\rho,\sigma}$ takes the value + for any model of $\mathscr{T}^*$ and is therefore in $\mathscr{T}^*$. Conversely, let P be a formula such that $P_{\rho,\sigma}$ is in $\mathscr{T}^*$. Then for any model R of $\mathscr{T}$ there exists a multirelation S such that RS is a model of $\mathscr{T}^*$, so that $P_{\rho,\sigma}(RS) = +$ and thus P(R) = +. Hence every model of $\mathscr{T}$ satisfies P, and so P is in $\mathscr{T}$. $\triangleleft$

6.3.2. *Let $\mathcal{T}$ be a theory and $\mathcal{T}^*$ an expansion of $\mathcal{T}$; let* R *be a multirelation assignable to $\mathcal{T}$ and* S *another multirelation. If* RS *is a model of $\mathcal{T}^*$, then* R *is a model of $\mathcal{T}$.*

In fact, let P be a formula of $\mathcal{T}$; be assumption, $P_{\rho,\sigma}$ is in $\mathcal{T}^*$. Since RS is a model of $\mathcal{T}^*$, we have $P_{\rho,\sigma}(RS) = +$, and so $P(R) = +$.

6.3.3. *Let $\mathcal{T}$ be a theory and $\mathcal{T}^*$ an expansion of $\mathcal{T}$. For any multirelation* R *with finite base which is a model of $\mathcal{T}$ there exists a multirelation* S *such that* RS *is a model of $\mathcal{T}^*$.*

▷ Suppose the contrary: let R be a model of $\mathcal{T}$ with finite base, such that for any S over the same base there is a formula P_S in $\mathcal{T}^*$ with $P_S(RS) = -$. Since there are only finitely many possible multirelations S, we may consider the conjunction P of all the formulas P_S. P is a formula of $\mathcal{T}^*$ and takes the value – for R and all S. Letting p be the cardinality of the base, we may identify the base with the set of natural numbers $1, \ldots, p$. With each pair (i, j), $1 \leqslant i < j \leqslant p$, we associate the formula $x^i \not\equiv x^j$; consider the formula

$$\underset{p+1}{\forall}\ x^1 \equiv x^{p+1} \bigvee \cdots \bigvee x^p \equiv x^{p+1}.$$

Let m be the arity of R. With each m-tuple of numbers $i_1, \ldots, i_m$, $1 \leqslant i_j \leqslant p$, we associate the formula $\rho x^{i_1} \ldots x^{i_m}$ or its negation, according as $R(i_1, \ldots, i_m) = +$ or $-$. Let Q denote the conjunction of all the above formulas.

The conjunction $P \bigwedge Q$ is an antithesis: its value is $-$ for any R′ assignable to ρ, any S′ assignable to σ and any elements $a_1, \ldots, a_p$ of the base of R′S′. We consider two cases: (1) The mapping of $1, \ldots, p$ onto $a_1, \ldots, a_p$ is an isomorphism of R onto a restriction of R′; then P takes the value $-$. (2) The above mapping is not an isomorphism; then Q takes the value $-$. Finally, the formula $Q' = \neg \underset{i}{\exists} \cdots \underset{p}{\exists} Q$ is deducible from P, which is a bound formula belonging to $\mathcal{T}^*$. Thus Q′, viewed as a formula in ρ alone is in $\mathcal{T}$ and therefore takes the value $+$ for the model R of $\mathcal{T}$; on the other hand, by construction its value is $-$. This contradiction completes the proof. ◁

6.3.4. *If $\mathcal{T}^*$ is a given expansion of $\mathcal{T}$ and* R *a model of $\mathcal{T}$, there exist multirelations* R′ *and* S′ *such that* R′ *is logically equivalent to* R *and* R′S′

is a model of $\mathcal{T}^*$ (Silver, reported by [CRA, 1960]; another proof has been given by A. Astier, 1970, unpublished).

▷ The theory $\mathcal{T}$ is obviously interpretable in $\mathcal{T}^*$, so that our assertion follows from 6.2.4. ◁

6.3.5. Let $\mathcal{T}$ be a theory all of whose models are finite. Then, by 6.3.2 and 6.3.3, any expansion of $\mathcal{T}$ is semantic. We shall now demonstrate the existence of theories with infinite models, even *only* infinite models, all of whose expansions are semantic.

6.3.5.1. *Any expansion of the theory consisting of predicate-free theses is semantic.*

▷ Let $\mathcal{T}$ be the theory of predicate-free bound theses, and let σ be a sequence of additional predicates. A theory $\mathcal{T}^*$ is an expansion of $\mathcal{T}$ to σ if and only if the only formulas of $\mathcal{T}^*$ in which σ is inactive are the theses. Examples are the theory consisting of all theses in σ, or the theory generated by the formula $\underset{x}{\forall}\sigma x$ (where σ consists of a single unary predicate; see 6.2). On the other hand, a theory is a semantic expansion of $\mathcal{T}$ to σ if and only if it has a model over any set.

Any semantic expansion is an expansion (6.3.1). This may also be verified directly: if a formula P has a model over every set and the predicates of P are inactive, then P is a thesis.

Conversely, let $\mathcal{T}^*$ be an expansion of $\mathcal{T}$. There are two possibilities: (1) there exists p such that $\mathcal{T}^*$ has no model of cardinality p. Then the bound formula with no active predicates stating that the number of elements of the base is not exactly p lies in the theory $\mathcal{T}^*$, but it is not a thesis: this contradicts the definition of an expansion. (2) $\mathcal{T}^*$ has models of arbitrary finite cardinality; then it has a model over any infinite base (see 2.2.4 and 4.4.5), so that $\mathcal{T}^*$ is a semantic expansion. ◁

6.3.5.2. *Any expansion of the theory of infinite sets is semantic.*

▷ Let $\mathcal{T}$ be the theory of infinite sets and $\mathcal{T}^*$ an expansion. For every p, the bound formula with no active predicate stating that the base contains at least p elements is in the expansion $\mathcal{T}^*$; thus $\mathcal{T}^*$ cannot have finite models. At the same time, $\mathcal{T}^*$ (like $\mathcal{T}$) is consistent, and therefore has a model, which must have an infinite base. Consequently, $\mathcal{T}^*$ has a model over any infinite set, by the restriction theorem 1.4.4 and the extension theorem 4.4.5. Thus $\mathcal{T}^*$ is a semantic expansion of $\mathcal{T}$. ◁

6.3.6. *The saturated theory of the chain of natural numbers has*
(1) *a saturated nonsemantic expansion;*
(2) *a finitely-axiomatizable nonsemantic expansion.*

▷ (1) Let I denote the chain and S the sum of natural numbers, $\mathcal{T}$ the saturated theory of I. Consider the saturated theory $\mathcal{T}^*$ of the birelation (I, S); this is an expansion of $\mathcal{T}$. Let I′ be the chain of natural numbers followed by the chain of integers; we know that I′ is logically equivalent to I (see 1.4, referring back to 1.1.6). Thus I′ is a model of $\mathcal{T}$; however, there is no relation S′ such that (I′, S′) is a model of $\mathcal{T}^*$, i.e., logically equivalent to (I, S) (see 1.4.2).

(2) Let $\mathcal{T}$ be as before (the saturated theory of I) and let $\mathcal{T}^*$ be the theory of all birelations which are (3, 6)-equivalent to (I, S). $\mathcal{T}^*$ is generated by the bound formulas of characteristic at most (3, 6) which are true for (I, S); since the number of all such formulas is finite (up to equideducibility), $\mathcal{T}^*$ is finitely-axiomatizable. Let I′ be as before. By 1.4.2, there is no relation S′ such that (I′, S′) is a model of $\mathcal{T}^*$. Thus $\mathcal{T}^*$ is not a semantic expansion of $\mathcal{T}$. All the axioms of I are accounted for, since their characteristic is (3, 4) (see 1.5.1). ◁

6.3.7. *Let ρ be a binary predicate and σ a ternary predicate. The theory of bound theses in ρ has a nonsemantic expansion in the predicates ρ, σ.*

▷ Let I, S be the chain and sum of natural numbers, $\mathcal{T}$ the theory of bound theses in the single predicate ρ, and $\mathcal{T}^*$ the theory in ρ, σ generated by the formulas expressing the following statements. The relation replacing ρ is either not a discrete chain with minimal element but no maximal element, or it is a chain of this type and together with the relation replacing σ it forms a birelation (2, 6)-equivalent to (I, S). Then $\mathcal{T}^*$ is not a semantic expansion of $\mathcal{T}$, for if I′ (the chain of natural numbers followed by all integers) is substituted for ρ there is no relation S′ such that (I′, S′) is a model of $\mathcal{T}^*$ (as follows from 1.4.2).

It remains to show that $\mathcal{T}^*$ is an expansion of $\mathcal{T}$. Let P be a formula in $\mathcal{T}^*$ in which σ is inactive, and R any binary relation. Now, if R is a discrete chain with minimal but no maximal element, then P(R, X) = +

for any ternary relation X with the same base. If R is a chain of this type, it is logically equivalent to I. By hypothesis, P(I, S) = +, so that P(I, X) = + for any ternary relation X over the natural numbers, in particular, for the relation X which is + for all triples of base elements. Letting U denote the ternary relation which is + for all triples of elements of |R|, we see that the birelations (R, U) and (I, X) are logically equivalent. Thus P(R, U) = + and so P(R, V) = + for any ternary relation V over |R|. Thus P is a thesis and it is the canonical transform of a formula of $\mathscr{T}$. ◁

6.3.8. *There exists a saturated finitely-axiomatizable theory which has a saturated finitely-axiomatizable nonsemantic expansion.*

▷ Let $\mathscr{T}$ be the saturated theory of the chain I of natural numbers, or of the logically equivalent chain I′ of the natural numbers followed by all the integers. Let A′ be the unary relation which is + for the natural numbers, − for the other elements, and let $\mathscr{T}^*$ be the saturated theory of (I′, A′) (see Chapter 1, Exercise 4.3). Then I is a model of $\mathscr{T}$, showing that $\mathscr{T}^*$ is not a semantic expansion of $\mathscr{T}$. ◁

6.4. Axiomatizable multirelations and theories

A theory is said to be *axiomatizable* if it admits a finitely-axiomatizable expansion. A multirelation R is said to be *axiomatizable* if the saturated theory of R is axiomatizable.

In particular, a multirelation R is axiomatizable if there exists a multirelation S with the same base such that RS is finitely-axiomatizable.

Examples. Any finitely-axiomatizable multirelation or theory is of course axiomatizable.

The succession relation C on the natural numbers is axiomatizable, for if I denotes the chain on the natural numbers, the birelation (C, I) is finitely-axiomatizable (see 1.5.2). We know, however, that C is not finitely-axiomatizable (1.5.3).

6.4.1. *Let $\mathscr{T}$ and $\mathscr{U}$ be two theories. If $\mathscr{T}$ is axiomatizable and $\mathscr{U}$ is interpretable in $\mathscr{T}$, then $\mathscr{U}$ is axiomatizable.*

▷ Let ρ be the predicates in $\mathscr{T}$; let $\mathscr{T}^*$ be a finitely-axiomatizable expansion of $\mathscr{T}$, with predicates ρ, σ. Consider the conjunction of a single axiom for $\mathscr{T}^*$ and the definitions of all the predicates τ of $\mathscr{U}$ in terms of ρ

(since the predicates are finite in number, so are these definitions). The result is a finitely-axiomatizable theory $\mathscr{U}^*$ in predicates ρ, σ, τ, which is moreover an expansion of $\mathscr{U}$. Indeed, any formula of $\mathscr{U}^*$ in which ρ, σ are inactive is necessarily deducible from the definitions and from a formula of $\mathscr{T}^*$ in which σ is inactive, hence also from a formula of $\mathscr{T}$. $\lhd$

6.4.2. *The sum of natural numbers is axiomatizable* (it is not finitely-axiomatizable; see 3.5.3).

$\rhd$ Let S denote the sum of natural numbers; we associate with S a predicate σ and let $\mathscr{T}$ denote the saturated theory of S. Consider the product P over the natural numbers and let ϖ be a corresponding predicate. It will suffice to construct an expansion $\mathscr{T}^*$ of $\mathscr{T}$ in the predicates σ, ϖ which is finitely-axiomatizable. This we now proceed to do, actually exhibiting an axiom system for $\mathscr{T}^*$.

We first take a finite set of formulas, with σ as the only active predicate, stating that the relation to be substituted for σ is a positive discrete sum (see 3.4 and 3.5). By 3.5.2, it will suffice to add a finite set of formulas which, together with the preceding ones, state that the sum is p-divisible for every natural number p (via deduction). We let $+$ denote addition and $\cdot$ multiplication (the reader should replace these, strictly speaking, by the predicates σ and ϖ), and consider the following four formulas:

$\forall_x x\cdot 0=0$, which is an abbreviation for $\forall_u\forall_x(u+u=u)\Rightarrow(x\cdot u=u)$;

$$\forall_x\forall_y x\cdot(y+1)=(x\cdot y)+x,$$

which is an abbreviation for

$$\forall_u\forall_v((u+u=u\wedge v\not\equiv u\wedge\forall_z(z\equiv u\vee v\leqslant z))\Rightarrow$$
$$\Rightarrow\forall_x\forall_y x\cdot(y+v)=(x\cdot y)+x),$$

where $v\leqslant z$ is an abbreviation for $\exists_t v+t=z$;

$$\forall_w\forall_{w'} w'=w+1\Rightarrow\forall_y(y<w'\Leftrightarrow(y<w\vee y\equiv w)),$$

where $y<w$ is an abbreviation for $\exists_t y+t=w\wedge y\neq w$;

$$\forall_x\forall_w w\neq 0\Rightarrow\exists_y\exists_t(x=(y\cdot w)+t\wedge t<w).$$

All the formulas

$$\underset{x}{\forall}\underset{y}{\exists}\, x=y+y \bigvee x=y+y+1$$

and

$$\underset{x}{\forall}\underset{y}{\exists}(x=y+y+y \bigvee x=y+y+1 \bigvee x=y+y+y+2),$$

and so on, are all deducible from the above axiom system (set w equal to 2, 3,..., p,... successively). Thus the sum is p-divisible for every natural number p. ◁

Problems. (1) Gödel's celebrated incompleteness theorem [GOD, 1931] implies that the saturated theory of the birelation (S, P) (sum and product over natural numbers) is not axiomatizable. Is there a multirelation N other than P, with the same base (natural numbers), such that SN is finitely-axiomatizable?

(2) Generalizing the above problem, we may ask: if M is axiomatizable, does there always exist a multirelation N such that MN is finitely-axiomatizable? In other words, if the saturated theory of M admits a finitely-axiomatizable expansion, does there exist a *saturated* finitely-axiomatizable expansion?

(3) Above we constructed a finitely-axiomatizable expansion of the saturated theory of the sum over natural numbers. Is this expansion semantic?

Note that if we enlarge the axiom system by adding the formula $\underset{x}{\forall}\underset{y}{\forall}\underset{z}{\exists}\varpi xyz$, i.e., we demand that the product associate an element $c=a\cdot b$ with each pair a, b, the expansion is not semantic. The proof is analogous to that in 3.5.4. Replace σ by the sum S over pairs (a, b), where a is a nonnegative rational number and b an integer (natural number if $a=0$, arbitrary integer if $a>0$). If the expansion is semantic, there exists a relation P over the pairs such that (S, P) satisfies the above axiom system. Then the result of multiplying (1, 0) by itself is such that

$$\begin{aligned}(1, 0)\cdot(1, 0)&=(1, 0)\cdot((1, -n)+(0, n))=\\&=(1, 0)\cdot(1, -n)+(n, 0)>(n, 0)\end{aligned}$$

for every natural number n; but this is impossible, for there is no pair following $(n, 0)$ for all n (private communication of A. Roberty).

Remark. Most authors define an axiomatizable theory in terms of the existence of a *decidable* axiom system; the notion of decidability (existence of a decision procedure) involves recursive function theory and "gödelization" of the set of formulas of a given arity. In the case of theories having only infinite models, Kleene's theorem [KLE, 1952], as sharpened by [CRA-VAU, 1958], proves that our notion is equivalent to the traditional axiomatizability. For theories having a finite model, our axiomatizability is stronger; one must then distinguish between strong and weak axiomatizability.

6.5. Free expansion

Let $\mathscr{T}$ be a theory in predicates ρ, and let σ be additional predicates. A theory $\mathscr{T}'$ in ρ, σ will be called a *free expansion* of $\mathscr{T}$ if the free formulas in which the σ are inactive and which are deducible from formulas of $\mathscr{T}'$ are precisely the canonical transforms of the free formulas deducible from formulas of $\mathscr{T}$. Equivalently, $\mathscr{T}'$ and the canonical expansion $\mathscr{T}_{\rho,\sigma}$ of $\mathscr{T}$ generate the same set of deducible free formulas with the predicates σ inactive.

6.5.1. *Any expansion of $\mathscr{T}$ is a free expansion of $\mathscr{T}$.*

On the other hand, *there exists a theory $\mathscr{T}$ having a free expansion which is not an expansion of $\mathscr{T}$*. It may even be required that the free expansion in question have the same predicates as $\mathscr{T}$.

▷ Let $\mathscr{T}$ be the set of bound theses in a single unary predicate ρ, and $\mathscr{T}'$ the theory consisting of all bound formulas deducible from $\underset{u}{\exists}\rho u$. Then $\mathscr{T}'$ is not an expansion of $\mathscr{T}$, for it contains $\underset{u}{\exists}\rho u$, which is not a formula of $\mathscr{T}$. Nevertheless, the free formulas deducible from $\mathscr{T}$ and $\mathscr{T}'$ are the same, so that each of these theories is a free expansion of the other. Indeed, let $\mathrm{P}(x, y, \ldots)$ be a free formula deducible from $\underset{u}{\exists}\rho u$, with free indices $x, y, \ldots$ If this formula is not in $\mathscr{T}$, it is not a thesis. Hence there exist a unary relation R and elements $a, b, \ldots \in |\mathrm{R}|$ such that $\mathrm{P}(\mathrm{R})(a, b, \ldots) = -$. Since P is free, this value is $-$ whatever the values taken by R on elements other than $a, b, \ldots$, in particular, if we add to the base an element u such that $\mathrm{R}(u) = +$. But then $\underset{u}{\exists}\rho u$ takes the value $+$, so that P cannot be deducible from it – contradiction. ◁

EXERCISES

1

(1) Let R be a μ-ary multirelation with base E, R′ and S′ two multirelations with the same base, of arities μ and ν. Suppose that every finite restriction of R is $\leqslant$R′. Then, for any finite subset F of E, there exist a ν-ary multirelation S_F with base F and an extension of $(R \mid F, S_F)$ which is isomorphic to (R′, S′). Show that there exists a multirelation S with base E such that (R, S) has an extension logically equivalent to (R′, S′). *Hint:* The multirelations S_F satisfy the assumptions of the coherence lemma (Volume 1, 3.1.3): for any subset G of F, the restriction of S_F to G is one of the multirelations S_G; then use 4.4.3.

(2) Let R be a μ-ary multirelation with base E and $\mathscr{A}$ a logical class of arity (μ, ν). Suppose that for any finite subset F of E there exist a multirelation S_F and an extension of $(R \mid F, S_F)$ which belongs to $\mathscr{A}$. Show that there exists a multirelation S with base E such that (R, S) has an extension in $\mathscr{A}$.

(3) In other words, for any multirelation R and $\mathscr{S}$-logical class $\mathscr{B}$ (see 6.1), if every finite restriction of R has an extension in $\mathscr{B}$, then R itself has an extension in $\mathscr{B}$. Hence deduce that the class of ordinals is not an $\mathscr{S}$-logical class.

2

Show that the following conditions are equivalent:

(1) There exists a theory which is stronger than $\mathscr{U}$ and interpretable in $\mathscr{T}$.

(2) There exists a theory weaker than $\mathscr{T}$ in which $\mathscr{U}$ is interpretable.

3

Let E be the set of all integers, 0 the singleton zero, C the succession relation, U the binary relation such that $U(0, 1) = +$ (and $-$ otherwise), B the "double succession" relation: $B(x, y) = +$ if and only if $y = x + 2$.

(1) Show that UB is interpretable in 0C, 0B is interpretable in UB, and 0B and 0C are logically equivalent. Thus $\mathscr{T}$, the saturated theory of 0C, and $\mathscr{U}$, the saturated theory of UB, are interpretable in one another.

(2) Show that C is not interpretable in UB. *Hint:* For all k, p, there is an integer a such that the mapping f defined by $f(a) = a + 2$, $f(a+1) = a + 1$, is a (k, p)-automorphism of UB but not a local isomorphism of C.

(3) Show that any birelation logically equivalent to 0C is isomorphic to a logical extension of 0C; the same is true for UB. *Hint:* All the n-tuples of integers are interpretable both in 0C and in UB.

(4) Suppose that there exist two logical operators $\mathscr{P}$, $\mathscr{Q}$ such that $\mathscr{P}$(UB) is logically equivalent to 0C and $\mathscr{Q}\mathscr{P}$(UB) = UB. Then, by 2.5.2, $\mathscr{P}$ maps the class of all birelations logically equivalent to UB onto the class of all birelations logically equivalent to 0C. Thus $\mathscr{P}$(UB) is a logical extension of 0C, and so $\mathscr{Q}$(0C) = UB. Finally, $\mathscr{P}$(UB) is isomorphic to 0C and interpretable in UB, hence also in 0C. Note that 0C is not interpretable in $\mathscr{P}$(UB) for, if it were, 0C would be interpretable in UB, contrary to part 2. Now compare this situation with Exercise 8, Chapter 1, to conclude that $\mathscr{T}$ and $\mathscr{U}$ do not possess models which are interpretable in one another (A. Roberty, 1970, unpublished).

4

Let $P(\rho, \sigma)$ be a formula in predicates ρ, σ, and R a relation of arity ρ for which all bound formulas deducible from $P(\rho, \sigma)$ in which σ is inactive take the value $+$.

(1) Let $P(\rho, \sigma)$ express the statement: ρ is a chain, there exist elements satisfying the unary predicate σ but no such element is minimal. Show that the set of bound formulas deducible from $P(\rho, \sigma)$ in which σ is inactive is simply the theory of all infinite chains. Letting R be the chain of natural numbers, show that there exists no relation S such that $P(R, S) = +$.

(2) By contrast, using 6.2.4, we note that there exist relations R′, S′, where R′ is logically equivalent to R, such that $P(R', S') = +$.

Problems. (1) Let P and R satisfy the conditions stipulated at the beginning of Exercise 4. Does there exist a formula $P^*(\rho, \sigma)$, such that any formula with σ inactive which is deducible from P* is also deducible from P and vice versa, and there exists a relation S satisfying $P^*(R, S) = +$?

(2) Does there exist a formula P* possessing the above properties for any relation R satisfying the above condition?

In the preceding example, let $P^*(\rho, \sigma)$ state that ρ is a chain, there exist elements satisfying σ but no such element is maximal or minimal.

If σ is of arity $\geqslant 2$, certain results of [KLE, 1952] and [CRA-VAU, 1958] make it reasonable to expect that the solution to Problem 1 is positive.

5

Let $\mathscr{U}$ be a finite sequence of symbols $\mathscr{P}$, $\mathscr{S}$. We define a *$\mathscr{U}$-logical* class inductively, as follows. If $\mathscr{A}$ is a $\mathscr{U}$-logical class, we call $\mathscr{B}$ an $\mathscr{P}\mathscr{U}$-logical class provided $R \in \mathscr{B}$ if and only if $(R, S) \in \mathscr{A}$ for any S; the definition of an $\mathscr{S}\mathscr{U}$-logical class is analogous.

Problem. For any $\mathscr{U}$ whose first symbol is $\mathscr{P}$, does there exist an $\mathscr{S}\mathscr{U}$-logical class which is not a $\mathscr{U}$-logical class (and conversely, with $\mathscr{P}$ and $\mathscr{S}$ interchanged)?

6

Let us use the general term *pseudo-logical classes* for all the $\mathscr{U}$-logical classes as defined in Exercise 5. Assuming the axiom of constructibility, prove that any two denumerable ordinals which belong to the same pseudo-logical class are identical [MAR, 1973].

Problem. Are any two denumerable relations belonging to the same pseudo-logical class necessarily isomorphic (assuming the axiom of constructibility)?

CHAPTER 7

ULTRAPRODUCT

7.1. Family of Multirelations, Ultrafilter, Induced Logical Equivalence Class; Ultraproduct and Ultrapower; Maximal Case

Let I be some set and $\mathscr{F}$ an ultrafilter on I. Consider a family of multirelations R_i of the same arity μ, where i ranges over I. A class $\mathscr{A}$ is said to be *induced* by $\mathscr{F}$ if the set of indices i such that $R_i \in \mathscr{A}$ is an element of $\mathscr{F}$; another way of saying this is that $R_i \in \mathscr{A}$ for *almost every* i (mod $\mathscr{F}$). It is evident that for any class $\mathscr{A}$ of μ-ary multirelations either $\mathscr{A}$ or its complement is induced, and also that any finite intersection of induced classes is an induced class (and therefore nonempty). By the compactness theorem (2.2.3), the intersection of all logical classes induced by $\mathscr{F}$ is not empty; in fact, it is a logical equivalence class, known as the *logical equivalence class induced by* $\mathscr{F}$.

Let A_i denote the base of R_i. We define a *function* to be any set f of pairs $(i, f(i))$, $i \in I$, such that $f(i)$ is an element of A_i (exactly one element for each i). Two functions f, g are said to be equivalent mod $\mathscr{F}$ if $f(i) = g(i)$ for almost every i.

Let B be a set of pairwise nonequivalent functions. The *ultraproduct of the multirelations* R_i *over* B (mod $\mathscr{F}$) is the μ-ary multirelation S with base B defined as follows. We illustrate the definition for an *m-ary* relation. For each m-tuple of functions $f_1, \ldots, f_m$ in B, we set $S(f_1, \ldots, f_m) = +$ or $-$ according as $R_i(f_1(i), \ldots, f_m(i)) = +$ for almost every i or $R_i(f_1(i), \ldots, f_m(i)) = -$ for almost every i.

Suppose that all the R_i coincide, say, with a multirelation R with base A, and let B contain all the constant functions ($f(i) = a$ for all i, where $a \in A$); if we now identify each $a \in A$ with the corresponding constant function, B becomes a superset of A and the ultraproduct over B is an extension of R to B, called an *ultrapower* of R.

Again, let $\{R_i\}$ be a family of multirelations, $i \in I$. Let k, p be two natural numbers and consider a finite set D of functions f. Two indices

$i, j \in \mathrm{I}$ are said to be (k, p)-*equivalent* for D if the mapping taking $f(i)$ onto $f(j)$ for each $f \in \mathrm{D}$ is a (k, p)-isomorphism of R_i onto R_j. This is an equivalence relation, and the number of equivalence classes is finite. Thus exactly one of these classes is an element of $\mathscr{F}$; it is said to be (k, p)-*associated* with D.

7.1.1. The ultraproduct is said to be *maximal* if the set B contains a representative of each equivalence class of functions (mod $\mathscr{F}$).

Let S *be a maximal ultraproduct of* $\{\mathrm{R}_i\}$. *Let* D *be a finite set of functions* f, k *and* p *two natural numbers, and* U *an element of* $\mathscr{F}$ (k, p)*-associated with* D. *Then, for any* $i \in \mathrm{U}$, *the mapping taking each* f *onto* $f(i)$ *is a* (k, p)*-isomorphism of* S *onto* R_i.

▷ First let $k=0$. There exists an element V in the ultrafilter such that $\mathrm{S}(f_1, \ldots, f_m) = \mathrm{R}_j(f_1(j), \ldots, f_m(j))$ for all $j \in \mathrm{V}$ and any $f_1, \ldots, f_m \in \mathrm{D}$. The intersection $\mathrm{U} \cap \mathrm{V}$ is not empty; let $j \in \mathrm{U} \cap \mathrm{V}$. Then we have both the preceding equality and

$$\mathrm{R}_i(f_1(i), \ldots, f_m(i)) = \mathrm{R}_j(f_1(j), \ldots, f_m(j));$$

hence $\mathrm{V} = \mathrm{U}$ and the mapping taking each f onto $f(i)$ is a local isomorphism of S onto R_i.

Now let $k \geqslant 1$, and assume that the assertion is true for all natural numbers $<k$ and all p; we shall prove it for k and p.

Consider functions $g_1, \ldots, g_q$ (in the base of the ultraproduct), where $q \leqslant p$, and let V be the element of the ultrafilter $(k-1, p-q)$-associated with the set $\mathrm{D} \cup \{g_1, \ldots, g_q\} = \bar{\mathrm{D}}$. For each $j \in \mathrm{V}$, the mapping taking each f onto $f(j)$ and each g onto $g(j)$ is by hypothesis a $(k-1, p-q)$-isomorphism of S onto R_j. Let $j \in \mathrm{U} \cap \mathrm{V}$. Then the mapping taking each $f(j)$ onto $f(i)$ is a (k, p)-isomorphism of R_j onto R_i. Hence there exist elements $x_1, \ldots, x_q$ in the base A_i of R_i such that the mapping taking each $f(j)$ onto $f(i)$ and each $g(j)$ onto the corresponding x is a $(k-1, p-q)$-isomorphism of R_j onto R_i. Hence the mapping taking each f onto $f(i)$ and each g onto the corresponding x is a $(k-1, p-q)$-isomorphism of S onto R_i.

Conversely, consider elements $x_1, \ldots, x_q \in \mathrm{A}_i$, $q \leqslant p$. For each $j \in \mathrm{U}$, we have corresponding elements $y_{j,1}, \ldots, y_{j,q} \in \mathrm{A}_j$ such that the mapping taking each $f(i)$ onto $f(j)$ (f ranging over D) and each x onto the corresponding y_j is a $(k-1, p-q)$-isomorphism of R_i onto R_j. Let $g_1, \ldots, g_q$ be functions with values $y_{j,1}, \ldots, y_{j,q}$ for each $j \in \mathrm{U}$ and arbitrary values for $j \notin \mathrm{U}$. Since S is a maximal ultraproduct, its base contains functions $g'_1, \ldots, g'_q$ equivalent to $g_1, \ldots, g_q$, respectively. Hence there exists $\mathrm{V} \in \mathscr{F}$ such that $g_l(j) = g'_l(j)$ for each $l = 1, \ldots, q$ and each $j \in \mathrm{V}$. Let $j \in \mathrm{U} \cap \mathrm{V}$; then the mapping taking each $f(i)$ onto $f(j)$ and each x onto the corresponding $g'(j)$ is a $(k-1, p-q)$-isomorphism of R_i onto R_j. Let W be the element of the ultrafilter $(k-1, p-q)$-associated with $\mathrm{D} \cup \{g'_1, \ldots, g'_q\}$. By hypothesis, for each $j \in \mathrm{W}$ the mapping taking each $f(j)$ onto f and each $g'(j)$ onto g' is a $(k-1, p-q)$-isomorphism of R_j onto S. Now let $j \in \mathrm{U} \cap \mathrm{V} \cap \mathrm{W}$; then the mapping taking each $f(i)$ onto f and each x onto the corresponding g' is a $(k-1, p-q)$-isomorphism of R_i onto S. ◁

7.1.2. *If the ultraproduct* S *is maximal, it belongs to the logical equivalence class induced by the ultrafilter* [ŁOS, 1955].

▷ Let k, p be two natural numbers and $\mathscr{A}$ the logical (k, p)-equivalence class induced by the ultrafilter in the sense of 7.1. This class is unique, since the (k, p)-equivalence classes are pairwise disjoint. Let U be the element of the ultrafilter defining $\mathscr{A}$, i.e., the set of all indices i such that $\mathrm{R}_i \in \mathscr{A}$. In other words, U is (k, p)-associated with the empty set. By 7.1.1, the ultraproduct is (k, p)-equivalent to the multirelations R_i in U, and so the ultraproduct belongs to $\mathscr{A}$. Hence it also belongs to the logical equivalence class which is the intersection of all classes $\mathscr{A}$ when k and p are varied. ◁

In particular: *If all multirelations in a family* $\{\mathrm{R}_i\}$ *belong to the same logical class* $\mathscr{A}$, *then the maximal ultraproduct of* $\{\mathrm{R}_i\}$ *belongs to* $\mathscr{A}$. *If the* R_i *are all logically equivalent, their maximal ultraproduct is logically equivalent to them.*

This is not true for an arbitrary ultraproduct. As an example, take the family of all multirelations R with infinite bases and a set B consisting of a single function. Then the ultraproduct is a multirelation whose base is a singleton, which cannot be logically equivalent to any R.

7.1.3. *Any maximal ultrapower of a multirelation* R *is a logical extension of* R.

▷ Recall that in an ultrapower all the multirelations R_i are identical to R; denote the base of R by E. The base of the ultrapower contains all the constant functions taking values in E. For any finite set D of constant functions, the set of all indices i is (k, p)-associated with D for any k, p (see 7.1). For any fixed i, the mapping taking $f(i)$ onto f for every constant function f is a (k, p)-isomorphism of R onto the ultrapower (this follows from 7.1.1); this is true for any finite set of constant functions f and any k, p. Now each constant function f is identified with its unique value $f(i)$; thus the identity mapping on any finite subset of E is a (k, p)-isomorphism of R onto the ultrapower for any k, p, and this meets the requirements of the definition of a logical extension (1.4). ◁

7.2. Logical equivalence implies the existence of isomorphic ultrapowers

Throughout the sequel, a cardinal is an ordinal α which is not equipollent to any ordinal $<\alpha$.

7.2.1. *Let α be an infinite cardinal, E_i $(i<\alpha)$ a family of sets of the same cardinal α. For each $i<\alpha$ there exists a subset F_i of E_i, of cardinal α, such that the sets F_i are pairwise disjoint.*

▷ We shall use the axiom of choice in the following equivalent form: every cardinal is equipollent to its square. Let h be a bijective mapping of α onto its square α^2. For each $i<\alpha$, let $j(i)$ denote the first term of $h(i)$ (recall that the latter is a pair of ordinals $<\alpha$). We first prove the existence of an injective mapping taking each ordinal $i<\alpha$ onto an element $f(i)\in E_{j(i)}$. Indeed, let $i_0<\alpha$ and suppose that for all $i<i_0$ we have already defined an injective mapping f taking each $i<i_0$ onto an element $f(i)\in E_{j(i)}$. Then the set $\{f(i)\colon i<i_0\}$ does not exhaust $E_{j(i_0)}$, for the latter has cardinal $\alpha>i_0$ and α is not equipollent to i_0. We may thus choose $f(i_0)$ to be an element of $E_{j(i_0)}$ distinct from all the $f(i)$ $(i<i_0)$ (this involves an application of the axiom of choice).

With the mapping f thus defined, we define F_j, $j<\alpha$, as the set of elements $f(i)$ such that $j(i)=j$. Then $F_j\subseteq E_j$ by construction; and since f is injective, F_j is of cardinal α for each j and the sets F_j are pairwise disjoint. ◁

7.2.2. Let I be an index set of cardinal α, and $\{R_i\}$, $\{S_i\}$ $(i\in I)$ two families of multirelations of the same arity with bases A_i, B_i, respectively, of cardinals $\leqslant 2^\alpha$ (the cardinal of the power set of α). We assume the truth of the generalized continuum hypothesis, in the form: there is no cardinal β such that $\alpha < \beta < 2^\alpha$. Let A denote the set of functions defined on I such that $f(i) \in A_i$ for each $i \in I$, and B the set defined similarly for the sets B_i; these sets are all of cardinal $\leqslant 2^\alpha$.

Then one of the following two conditions is satisfied:

(1) *There exists a bound formula* P *such that the set of all i in* I *satisfying* $P(R_i) = -$ *or* $P(S_i) = +$ *is of cardinal* $< \alpha$.

(2) *There exist a mapping a of* 2^α *onto* A *and a mapping b of* 2^α *onto* B *such that, for any logical formula* P *and any substitution of ordinals* $x, y, \ldots < 2^\alpha$ *for the free indices of* P, *there exists a subset* H_P *of* I, *of cardinal* α, *such that for all* $i \in H_P$

$$P(R_i)\,(a(x)\,(i), a(y)\,(i), \ldots) = -$$

or

$$P(S_i)\,(b(x)\,(i), b(y)\,(i), \ldots) = +.$$

▷ Assuming that condition 1 fails to hold, we shall prove that condition 2 is satisfied. Using the axiom of choice, we may well-order the power set of α in correspondence with the smallest possible ordinal, which we denote by 2^α. Call an ordinal $< 2^\alpha$ even or odd according as its Cantor development in decreasing powers of ω ends in an even or odd natural number. Let a' be an arbitrary mapping of the even ordinals $< 2^\alpha$ onto A and b' an arbitrary mapping of the odd ordinals $< 2^\alpha$ onto B; we are going to complete these mappings so as to obtain the required mappings a and b.

Let x_0 be an ordinal $< 2^\alpha$. By the GCH (generalized continuum hypothesis), the cardinal of x_0 is $\leqslant \alpha$. To apply transfinite induction, we replace 2^α by x_0 in condition 2: the mappings a and b are assumed to be defined on x_0, mapping it onto subsets of A and B, and ordinals $x, y, \ldots < x_0$ are substituted for the free indices of P. If $x_0 = 0$, the formulas indicated in condition 2 are necessarily devoid of free indices, and so they are bound; condition 2 is then simply the negation of condition 1, which we have already assumed to hold. Now let $x_0 \neq 0$ and suppose that condition 2 holds for x_0. We must now extend this to ordinals $x, y, \ldots \leqslant x_0$ (i.e., go from x_0 to its immediate successor). The induction step for a limit ordinal is obvious, since for each ordinal a and b are extensions of the corresponding mappings for smaller ordinals and each formula involves finitely many free indices.

To fix ideas, let x_0 be even; then $a(x_0)=a'(x_0)$ is given, and we must construct $b(x_0)$; this is a function associating with each $i \in \mathrm{I}$ an element $b(x_0)(i) \in \mathrm{B}_i$, in such a way that condition 2 holds for $x, y, \ldots \leqslant x_0$. Let $\mathscr{P}$ denote the set of logical formulas P in which the ordinals $\leqslant x_0$ are substituted for the free indices. If x_0 is infinite, the cardinal of $\mathscr{P}$ is at most that of x_0; in any case, it is always $\leqslant \alpha$. For any formula P in which at least one free index is replaced by x_0 and the others by $x, y, \ldots < x_0$, we have one of the following two cases: (a)

$$\mathrm{P}(\mathrm{R}_i)(a(x_0)(i), a(x)(i), a(y)(i), \ldots) = -$$

for a set of indices i of cardinal α, and condition 2 is satisfied for any definition of $b(x_0)$; or (b) the above equality holds only for a set of indices of cardinal $< \alpha$. Then the same is true of the equality

$$\underset{x_0}{\exists}\, \mathrm{P}(\mathrm{R}_i)(a(x)(i), a(y)(i), \ldots) = -.$$

By the induction hypothesis and the fact that the indices remaining free in $\underset{x_0}{\exists}\, \mathrm{P}$ are replaced by ordinals $< x_0$, there exists a set H'_P of indices i, of cardinal α, such that

$$\underset{x_0}{\exists}\, \mathrm{P}(\mathrm{S}_i)(b(x)(i), b(y)(i), \ldots) = +.$$

We now use 7.2.1: since the set $\mathscr{P}$ of formulas P is of cardinal $\leqslant \alpha$, we can define subsets $\mathrm{H}_\mathrm{P} \subseteq \mathrm{H}'_\mathrm{P}$, of cardinal α, which are pairwise disjoint for distinct formulas P. We can now define $b(x_0)$. If i is in some H_P, this uniquely determines P, and we let $b(x_0)(i)$ be an element of B_i such that

$$\mathrm{P}(\mathrm{S}_i)(b(x_0)(i), b(x)(i), b(y)(i), \ldots) = +.$$

If i is not in any H_P, we let $b(x_0)(i)$ be an arbitrary element of B_i. Since each H_P is of cardinal α, condition 2 is satisfied for all formulas P. A similar argument holds if x_0 is odd (with R_i and S_i interchanged). $\lhd$

7.2.3. With the same notation as before, condition 7.2.2(2) implies the following:

There exist ultrafilters $\mathscr{U}$ and $\mathscr{V}$ on I, *such that the maximal ultraproduct of* $\{\mathrm{R}_i\}$ *modulo $\mathscr{U}$ and the maximal ultraproduct of* $\{\mathrm{S}_i\}$ *modulo $\mathscr{V}$ are isomorphic. Moreover, we may assume that all the elements of $\mathscr{U}$ and $\mathscr{V}$ are subsets of* I *equipollent to* I.

▷ For each formula P with ordinals $x, y, \ldots < 2^\alpha$ substituted for the free indices, let U_P be the set of $i \in I$ such that

$$P(R_i)(a(x)(i), a(y)(i), \ldots) = +$$

and V_P the set of $i \in I$ such that

$$P(S_i)(b(x)(i), b(y)(i), \ldots) = +.$$

It is readily seen that $U_{\neg P}$ is the complement of U_P, $U_{P \wedge Q}$ is the intersection of U_P and U_Q and $U_{P \vee Q}$ their union, for any formulas P and Q; the same holds for the sets V. Let $\mathscr{V}_0$ be the set of all V_P such that $U_{\neg P}$ is of cardinal $<\alpha$. By 7.2.2, every element of $\mathscr{V}_0$ is of cardinal α. Moreover, the intersection of any two elements of $\mathscr{V}_0$ is an element of $\mathscr{V}_0$, because $V_{P \wedge Q}$ is the intersection of V_P and V_Q for all P and Q, and if both $U_{\neg P}$ and $U_{\neg Q}$ are of cardinal $<\alpha$ then the same is true of their union $U_{\neg(P \wedge Q)}$. The elements of $\mathscr{V}_0$, together with all subsets of I whose complements are subsets of cardinal $<\alpha$, have the finite intersection property; thus these sets, together with their intersections and supersets, constitute a filter on I; let $\mathscr{V}$ denote an ultrafilter refining this filter.

Let $\mathscr{U}_0$ be the set of all U_P such that $V_P \in \mathscr{V}$. Every element of $\mathscr{U}_0$ is of cardinal α. Indeed, if U_P is of cardinal $<\alpha$, then $V_{\neg P}$ is in $\mathscr{V}_0$, hence also in $\mathscr{V}$, and its complement V_P is not in $\mathscr{V}$ – contradiction. Moreover, the intersection of any two elements of $\mathscr{U}_0$ is an element of $\mathscr{U}_0$. Thus the elements of $\mathscr{U}_0$, together with all subsets of I whose complements are subsets of cardinal $<\alpha$, generate a filter on I, and we let $\mathscr{U}$ denote a corresponding ultrafilter.

For any formula P in which ordinals $<2^\alpha$ are substituted for the free indices, $U_P \in \mathscr{U}$ if and only if $V_P \in \mathscr{V}$. Indeed, if $V_P \in \mathscr{V}$, then $U_P \in \mathscr{U}_0 \subseteq \mathscr{U}$. On the other hand, if $V_P \notin \mathscr{V}$, then $V_{\neg P} \in \mathscr{V}$, so that $U_{\neg P} \in \mathscr{U}$; hence $U_P \notin \mathscr{U}$.

Let x, y be two ordinals $<2^\alpha$. We claim that $a(x)(i) = a(y)(i)$ for almost all $i \bmod \mathscr{U}$ if and only if $b(x)(i) = b(y)(i)$ for almost all $i \bmod \mathscr{V}$. Indeed, we need only let P be the free formula $x \equiv y$; it is immediate that U_P and V_P are precisely the sets of indices i satisfying these equalities.

The mapping taking $a(x)$ onto $b(x)$ for each ordinal $x < 2^\alpha$ takes any two functions $a(x)$, $a(x')$ which are equivalent modulo $\mathscr{U}$ onto functions $b(x)$, $b(x')$ which are equivalent modulo $\mathscr{V}$, and conversely (where x and x' are ordinals). Moreover, recall that the ranges of the functions a and b exhaust the entire sets A and B (see 7.2.2). Thus the above mapping induces on the equivalence classes of functions a bijective mapping of the ultraproduct of $\{R_i\}$ modulo $\mathscr{U}$ onto the ultraproduct of $\{S_i\}$ modulo $\mathscr{V}$. This mapping is an isomorphism. Indeed, let R and S denote these ultraproducts, and m their common arity. Given ordinals $x_1, \ldots, x_m < 2^\alpha$, determining elements $a(x_1), \ldots, a(x_m)$ of A and corresponding elements $b(x_1), \ldots, b(x_m)$ of B (up to equivalence modulo $\mathscr{U}$ and $\mathscr{V}$, respectively), we have

$$R(a(x_1), \ldots, a(x_m)) = + \quad \text{or} \quad -$$

according as $\mathscr{U}$ contains the set of indices i such that

$$R_i(a(x_1)(i), \ldots, a(x_m)(i)) = +$$

or the set of indices such that

$$R_i(a(x_1)(i), \ldots, a(x_m)(i)) = -.$$

Similarly,

$$S(b(x_1), \ldots, b(x_m)) = + \quad \text{or} \quad -$$

according as $\mathscr{V}$ contains the set of i such that

$$S_i(b(x_1)(i), \ldots, b(x_m)(i)) = +$$

or the set of i such that

$$S_i(b(x_1)(i), \ldots, b(x_m)(i)) = -.$$

Now these are precisely the sets U_P and V_P determined by the free formula P that transforms each m-ary relation into itself; thus the truth values in question are either both + or both −. ◁

7.2.4. It follows from the foregoing arguments that either condition 7.2.2(1) or condition 7.2.3 is satisfied. We now show that these conditions are incompatible: *each is equivalent to the negation of the other.*

▷ Suppose that 7.2.2(1) is satisfied: there exists a bound formula P such that $P(R_i) = +$ and $P(S_i) = -$ for all i outside a set of cardinal $< \alpha$. Suppose that the condition of 7.2.3 holds: the above equalities are valid for almost all $i \bmod \mathscr{U}$ and almost all $i \bmod \mathscr{V}$. It follows that P takes the value + for the maximal ultraproduct of $\{R_i\}$ modulo $\mathscr{U}$, and − for the maximal ultraproduct of $\{S_i\}$ modulo $\mathscr{V}$ (see 7.1.2). Thus the ultraproducts are not logically equivalent, and *a fortiori* not isomorphic. ◁

7.2.5. THEOREM. *Two multirelations are logically equivalent if and only if they have isomorphic maximal ultrapowers* (assuming the axiom of choice and GCH [KEI, 1961]; a proof avoiding GCH has been obtained by [SHE, 1971]).

▷ Let R and S be two multirelations. If they possess isomorphic maximal ultrapowers, the latter are logically equivalent. Hence, by 7.1.2 or 7.1.3, R and S are logically equivalent.

Conversely, let R and S be logically equivalent. Let I be a set whose cardinal α is so large that the bases of R and S are of cardinal $\leqslant 2^\alpha$. Let $R_i = R$, $S_i = S$ for all $i \in I$. Since R and S are logically equivalent, condition 7.2.2(1) cannot hold, and hence, as shown above, condition 7.2.3 must hold; in this case, the isomorphic ultraproducts are ultrapowers. ◁

7.3. CHARACTERIZATION OF LOGICAL CLASSES

7.3.1. *Let $\mathscr{A}$ and $\mathscr{B}$ be disjoint classes of multirelations, each closed under isomorphism and formation of maximal ultraproducts. Then the logical closures of $\mathscr{A}$ and $\mathscr{B}$ are disjoint* (the closure of $\mathscr{A}$ is the intersection of all logical classes containing $\mathscr{A}$; see 2.1).

▷ (Assuming AC and GCH.) Suppose the assertion false. Then there are a sequence of multirelations $R_i \in \mathscr{A}$ and a sequence $S_i \in \mathscr{B}$ $(i = 1, 2, \ldots)$ which are logically convergent to the same multirelation T (i.e., any logical class containing T contains R_i and S_i for all sufficiently large i; see 2.1). Consider the filter on the natural numbers consisting of all complements of finite sets, refine it to an ultrafilter, and let R be the corresponding maximal ultraproduct of $\{R_i\}$; construct S similarly for $\{S_i\}$. It is evident that $R \in \mathscr{A}$, $S \in \mathscr{B}$, and R and S are both logically equivalent to T. By 7.2.5, R and S possess isomorphic ultrapowers, which must be in both $\mathscr{A}$ and $\mathscr{B}$ – contradiction. ◁

7.3.2. *$\mathscr{A}$ is a logical class if and only if both $\mathscr{A}$ and its complement are closed under isomorphism and formation of maximal ultraproducts.*

▷ (Assuming AC and GCH.) If $\mathscr{A}$ is a logical class, so is its complement, and both are closed under isomorphism and formation of ultraproducts (see 7.1.2). Conversely, let $\mathscr{A}$ and its complement $\neg\mathscr{A}$ be closed under isomorphism and formation of maximal ultraproducts. By 7.3.1, their closures are disjoint, and therefore coincide with $\mathscr{A}$ and $\neg\mathscr{A}$, respectively. Each of these classes is therefore an intersection of logical classes, say $\mathscr{A}=\bigcap \mathscr{A}_i$ and $\neg\mathscr{A}=\bigcap \mathscr{A}'_i$. Now the intersection of the classes $\mathscr{A}_i \cap \mathscr{A}'_i$ is empty, and so $\mathscr{A}$ is the intersection of a finite subset of the $\mathscr{A}_i$. Hence $\mathscr{A}$ is a logical class. ◁

7.3.3. *Let $\mathscr{A}$ and $\mathscr{B}$ be disjoint classes, $\mathscr{A}$ closed under isomorphism and formation of maximal ultraproducts and $\mathscr{B}$ closed under isomorphism and formation of maximal ultrapowers. Then the logical closure of $\mathscr{A}$ is disjoint from $\mathscr{B}$.*

▷ (Assuming AC and GCH.) Suppose there exists a multirelation R both in the closure of $\mathscr{A}$ and in $\mathscr{B}$. There exists a sequence of multirelations R_i in $\mathscr{A}$ which converges logically to R (see 2.1). Let S be a maximal ultraproduct of $\{R_i\}$; then $S \in \mathscr{A}$ and S is logically equivalent to R. Let R^* be a maximal ultrapower of R and S^* a maximal ultrapower of S, such that R^* and S^* are isomorphic. Then $R^* \in \mathscr{B}$; therefore $S^* \in \mathscr{B}$ and $S^* \in \mathscr{A}$ – contradiction. ◁

7.3.4. *$\mathscr{A}$ is an intersection of logical classes if and only if it is closed under isomorphism and formation of maximal ultraproducts and its complement is closed under formation of maximal ultrapowers.*

▷ (Assuming AC and GCH.) If $\mathscr{A}$ is an intersection of logical classes, it is closed under isomorphism and formation of ultraproducts (7.1.2). The complement $\neg\mathscr{A}$ is a union of logical classes; hence, if $R \in \neg\mathscr{A}$, any maximal ultrapower of R is again in $\neg\mathscr{A}$, by 7.1.2 or 7.1.3. Conversely, suppose that $\mathscr{A}$ is closed under isomorphism and formation of maximal ultraproducts and $\neg\mathscr{A}$ under formation of maximal ultrapowers. By 7.3.3, the logical closure of $\mathscr{A}$ is disjoint from $\neg\mathscr{A}$ and therefore coincides with $\mathscr{A}$; hence $\mathscr{A}$ is an intersection of logical classes. ◁

7.4. Normal ultraproduct; definitions and examples

Consider a family of multirelations R_i ($i \in I$), an ultrafilter $\mathscr{F}$ on I and a set B of functions which are pairwise nonequivalent mod $\mathscr{F}$. The corresponding ultraproduct is said to be *normal* if the following condition holds:

For any finite set C of functions $f \in B$, any finite set D of functions g (not necessarily in B) and any natural numbers k, p, there exist $U \in \mathscr{F}$ and, for each $g \in D$, a function $g' \in B$, such that for each $i \in U$ the mapping which fixes all $f(i)$, $f \in C$, and takes each $g(i)$ onto $g'(i)$, $g \in D$, is a (k, p)-automorphism of R_i.

We call g' the *substitute* of g.

7.4.1. *The maximal ultraproduct is normal.*

▷ For each g, let the substitute be a function $g' \in B$ which is equivalent to g mod $\mathscr{F}$. There exists an element $U \in \mathscr{F}$ such that $g(i) = g'(i)$ for every $i \in U$; for any such index i, the mapping defined above fixes all the elements $f(i)$ and $g(i)$, and it is a (k, p)-automorphism of R_i for all k, p. ◁

7.4.2. *The constant ultrapower is normal.*

Let $R_i = R$ for all $i \in I$, and let B be the set of constant functions f, i.e., $f(i) = a$ (where $a \in |R|$) for each $i \in I$. Let C be a finite set of constant functions f, D a finite set of arbitrary functions g, and k, p two natural numbers. Let us say that two indices i, j are equivalent if the mapping taking $f(i)$ onto $f(j) = f(i)$, $f \in C$, and $g(i)$ onto $g(j)$, $g \in D$, is a (k, p)-automorphism of R. The number of equivalence classes is finite (1.2.1), and so exactly one of them is an element U of the ultrafilter. Let $i_0 \in U$, and define the substitute of each function $g \in D$ to be the constant function g' such that $g'(i) = g(i_0)$ for each $i \in I$. Then, for any $i \in U$, the mapping fixing each $f(i)$ and taking each $g(i)$ onto $g'(i) = g(i_0)$ is a (k, p)-automorphism of R. ◁

7.4.3. *Let* R *be a multirelation with base* E *and* S *a logical restriction of* R *to a subset* F *of* E. *Let* $R_i = R$ *for all* $i \in I$, *and let* B *be the set of constant functions with values in* F. *Then the ultraproduct of* $\{R_i\}$ *over* B *is normal (and isomorphic to* S*).*

▷ The proof is analogous to that of 7.4.2: one associates with each $g(i_0)$ a value $g'(i) = g'(i_0) \in F$, such that the mapping fixing the functions f and taking $g(i_0)$ onto $g'(i_0)$ for each g is a (k, p)-automorphism of R. This is possible, for if r is the number of functions g the identity mapping on the set of functions f is, among other things, a $(k+1, p+r)$-isomorphism of R onto S. ◁

7.4.4. *Example: the chain of natural numbers and the chain of natural numbers followed by the integers.*

▷ Let R denote the chain of natural numbers; let I be the set of natural numbers and $\mathscr{F}$ a nontrivial ultrafilter on I (i.e., an ultrafilter containing the complements of finite sets). Let $R_i = R$ for all $i \in I$, and let B be the set of constant functions, augmented by the following functions: for each constant (natural number) a, the function f such that $f(i) = i + a$, $i \in I$, and the function defined by $f(i) = 0$ for $i < a$, $f(i) = i - a$ for $i \geqslant a$. The ultrapower defined by these conditions is a chain S isomorphic to the chain of natural numbers followed by the integers. Indeed, for any two functions f_1, f_2 in B, we have $S(f_1, f_2) = +$ or $-$ according as $f_1(i) \leqslant f_2(i)$ or $f_1(i) > f_2(i)$ for all sufficiently large i.

We claim that this ultrapower is normal. Let C be a finite set of functions $f \in B$, D a finite set of arbitrary functions g, and k, p two natural numbers. There exists an element $U \in \mathscr{F}$ satisfying the following conditions:

(1) The order of the elements $f(i)$ and $g(i)$ is the same for each $i \in U$.

(2) Let h stand for any one of the functions f or g; then $h(i)$ is either constant and $< (p+1)^k$ for each $i \in I$, or $\geqslant (p+1)^k$ (and not necessarily constant) for each $i \in U$.

(3) Let h_1, h_2 stand for any two of the functions f or g; then $h_2(i) - h_1(i)$ is either constant and of absolute value $< (p+1)^k$ for each $i \in U$, or of the same sign and $\geqslant (p+1)^k$ for each $i \in U$.

(4) If a function g is constant or has the form (i + constant) for almost all i, then g has this form for all $i \in U$.

This follows from the fact that whenever the natural numbers i are partitioned into finitely many sets, one of the latter is an element of the ultrafilter.

By 1.1.6, it suffices to replace each $g \in D$ by a substitute $g' \in B$ such that, for almost every $i \in I$, the corresponding intervals defined by the functions f and g, on the one hand, and by the functions f and g', on the other, are either both of the same cardinality $< (p+1)^k$ or both of cardinalities $\geqslant (p+1)^k$. We construct the functions g' as follows.

First consider the functions $g \in D$ whose restriction to U is either a constant or a function (i + constant), and replace each of them by the function $g' \in B$ defined by $g'(i) = g(i)$ on U. Now divide the other functions

$g \in \mathrm{D}$ into two groups. The first group will contain all $g \in \mathrm{D}$ whose values on U lie between the greatest of the constant functions just considered, h say, and the smallest of the above functions (i + constant). Let $g_1, \ldots, g_r$ be the functions in this group, in increasing order of their values on U. We replace them by constant functions $g'_1, \ldots, g'_r$ defined as follows:

$$g'_1 = h + (p+1)^k;$$

$$g'_s - g'_{s-1} = \begin{cases} g_s - g_{s-1}, & \text{if} \quad g_s - g_{s-1} < (p+1)^k \quad \text{on} \quad \mathrm{U}, \\ (p+1)^k, & \text{otherwise} \end{cases}$$
$$(s = 2, \ldots, r).$$

The second group of functions will contain all $g \in \mathrm{D}$ exceeding the greatest of the above functions (i + constant), h^* say. Let $g_{r+1}, \ldots$ be the functions of this group, in increasing order of their values on U. We replace them by functions $g'_{r+1}, g'_{r+2}, \ldots,$ of the form (i + constant), defined as follows:

$$g'_{r+1}(i) = h^*(i) + (p+1)^k;$$

$$g'_{r+t} - g'_{r+t-1} = \begin{cases} g_{r+t} - g_{r+t-1}, \text{if } g_{r+t} - g_{r+t-1} < (p+1)^k \text{ on } U, \\ (p+1)^k, \quad \text{otherwise} \end{cases}$$
$$(t = 2, 3, \ldots). \triangleleft$$

Recall that the chain I′ of natural numbers followed by the integers is a logical extension of the chain I of natural numbers. By 7.4.3, I may be obtained as a suitable normal ultraproduct of multirelations identical with I′; by 7.4.4, I′ may be obtained in the same way from I. This is a particular case of 7.5.6 below.

7.5. Normal ultraproducts and logical equivalence

7.5.1. *Let* S *be a normal ultraproduct of* $\{\mathrm{R}_i\}$. *Let* C *be a finite set of functions* f *and* k, p *two natural numbers. Let* U *be an element of the ultrafilter* $\mathscr{F}$, (k, p)*-associated with* C. *Then, for each* $i \in \mathrm{U}$, *the mapping taking* f *onto* $f(i)$ *for all* $f \in \mathrm{C}$ *is a* (k, p)*-isomorphism of* S *onto* R_i.

$\triangleright$ The proof is analogous to that in 7.1.1, slightly modified at the end as follows. Since S is a normal ultraproduct, there exist substitutes $g'_1, \ldots, g'_q$ for $g_1, \ldots, g_q$ (in the base of S) and an element $\mathrm{V} \in \mathscr{F}$ such that for every $j \in \mathrm{V}$ the mapping fixing each $f(j)$, $f \in \mathrm{C}$, and mapping each $g(j)$ onto $g'(j)$ is a $(k-1, p-q)$-automorphism of R_j. $\triangleleft$

7.5.2. *If the ultraproduct is normal, it belongs to the family of logical classes induced by the ultrafilter, hence to the logical equivalence class induced by the ultrafilter.*

The proof is the same as in 7.1.2. Consequently, if all the multirelations R_i are in a logical class $\mathscr{A}$, any normal ultraproduct of $\{R_i\}$ is also in $\mathscr{A}$.

7.5.3. *A normal ultrapower is a logical extension of the original multirelation* (same proof as in 7.1.3).

7.5.4. *Any normal ultraproduct of multirelations* $\{R_i\}$ *is a logical restriction of the maximal ultraproduct of* $\{R_i\}$ *modulo the same ultrafilter.*

$\triangleright$ Let k, p be natural numbers and C a finite set of functions f in the base of the normal ultraproduct S. By 7.5.1, for almost every i the mapping taking each f onto $f(i)$, $f \in C$, is a (k, p)-isomorphism of S onto R_i. Complete the base of S by adding functions in such a way that each equivalence class of functions (modulo the ultrafilter) is represented. The result is a maximal ultraproduct T. By 7.1.1, for almost every i the mapping taking f onto $f(i)$, $f \in C$, is a (k, p)-isomorphism of T onto R_i. Finally, the identity mapping on C is a (k, p)-isomorphism of S onto T. $\triangleleft$

7.5.5. *Let* S *be a normal ultraproduct of* $\{R_i\}$; *then any logical restriction of* S *is a normal ultraproduct of* $\{R_i\}$.

$\triangleright$ For each $i \in I$, let E_i be the base of R_i, and let B be the set of functions which is the base of the ultraproduct S; let B′ be a subset of B, determining a logical restriction $S' = S \mid B'$ of S. Let C be a finite set of functions $f \in B'$ and D a finite set of arbitrary functions g. By assumption, there exist an element U of the ultrafilter and, for each $g \in D$, a substitute $g' \in B$, such that for each $i \in U$ the mapping fixing all $f(i)$, $f \in C$, and mapping $g(i)$ onto $g'(i)$, $g \in D$, is a (k, p)-automorphism of R_i. Let V be the element of the ultrafilter (k, p)-associated with $C \cup \{g' : g \in D\}$. By 7.5.1, for each $i \in V$ the mapping taking each f onto $f(i)$, $f \in C$, and each g' onto $g'(i)$, $g \in D$, is a (k, p)-isomorphism of S onto R_i. Hence, for each $i \in U \cap V$, the mapping taking each f onto $f(i)$ and each g' onto $g(i)$ is a (k, p)-isomorphism of S onto R_i.

Since S′ is a logical restriction of S, it follows that for each g' there exists a function $g'' \in B'$ such that the mapping fixing each f and taking each g' onto g'' is a (k, p)-isomorphism of S onto S′, hence also a (k, p)-automorphism of S. Let W be the element of the ultrafilter (k, p)-associated with $C \cup \{g'' : g \in D\}$. Now let $i \in U \cap V \cap W$ and use 7.5.1; then the mapping fixing each $f(i)$, $f \in C$, and taking each $g(i)$ onto $g''(i)$, $g \in D$, is a (k, p)-automorphism of R_i. ◁

By 7.4.1, 7.5.4 and 7.5.5 above, *an ultraproduct is normal if and only if it is a logical restriction of a maximal ultraproduct.*

7.5.6. (1) *Any multirelation logically equivalent to* R *is isomorphic to a normal ultraproduct of multirelations* $R_i = R$.

(2) *Any logical extension of* R *is isomorphic to a normal ultrapower of* R.

▷ (Assuming AC in the general case, and the ultrafilter axiom in the case of a logical extension with denumerable base.) We first prove part 2. Let E be the base of R and $E^* \supseteq E$ the base of a logical extension R^* of R. As index set we take the set of all triples $i = (F, k, p)$, where F ranges over the finite subsets of E^* and k, p are arbitrary natural numbers. With each index (F, k, p) we associate a (k, p)-isomorphism t_i of R^* onto R, with domain F, whose restriction to $F \cap E$ is the identity mapping (this involves an application of the axiom of choice if E^* is nondenumerable). With each element $a \in E^*$ we associate the function f_a defined as follows: if $a \in F$, we set $f_a(i) = t_i(a)$; if $a \in E$, we set $f_a(i) = a$; in all other cases $f_a(i)$ is an arbitrary element of E. Set $i' \geqslant i$, where $i' = (F', k', p')$, if $F' \supseteq F$, $k' \geqslant k$ and $p' \geqslant p$. With each i we associate the set of indices greater than i; the intersection of any two of these sets contains another set of the same kind, and so they generate a filter on the set of indices; let $\mathscr{F}$ be a corresponding ultrafilter.

We claim that with this ultrafilter the functions f_a define a normal ultrapower S of R with index set $\{i\}$; the mapping taking each $a \in E^*$ onto f_a is an isomorphism of R^* onto S.

We first show that if a, b are distinct elements of E^*, then f_a and f_b are not equivalent mod $\mathscr{F}$. Indeed, let $F = \{a, b\}$, $i = (F, 0, 0)$. For each $j \geqslant i$, and therefore for almost every index, we have $t_j(a) \neq t_j(b)$ (since t_j is injective), and so

$$f_a(j) = t_j(a) \neq f_b(j) = t_j(b).$$

We give the details for the case of m-ary relations. Let $a_1, \ldots, a_m \in E^*$. We claim that $S(f_{a_1}, \ldots, f_{a_m}) = R^*(a_1, \ldots, a_m)$. Indeed, let $F = \{a_1, \ldots, a_m\}$, $i = (F, 0, 0)$. Then, for each index $j \geqslant i$, hence for almost every index, we have

$$R(t_j(a_1),\dots, t_j(a_m))=R^*(a_1,\dots, a_m)$$

since t_j is a local isomorphism of R* onto R. Now $t_j(a)=f_a(j)$ for all $a_1,\dots, a_m$; finally, by the definition of the ultrapower S,

$$S(f_{a_1},\dots,f_{a_m})=R(f_{a_1}(j),\dots,f_{a_m}(j))$$

for almost every index j.

It remains to show that S is a normal ultrapower. Consider a finite set F of elements $a\in E^*$ and let C be the corresponding set of functions f_a; let D be a finite set of arbitrary functions g on the indices i, with values in E; let q be the cardinality of D and k, p two natural numbers. Let U be an element of $\mathscr{F}$, (k, p)-associated with $C\cup D$ (see 7.1), consisting of indices greater than $(F, k+1, p+q)$. Let $i_0\in U$ and let $t_0=t_{i_0}$ be the local isomorphism of R* onto R associated with i_0; t_0 is (at least) a $(k+1, p+q)$-isomorphism. Hence t_0 has an extension t_0^* which is a (k, p)-isomorphism of R* onto R, mapping a new set D′ of q elements $b\in E^*$ onto the elements $g(i_0)$. Let F* denote the union of F and all the b's, and let V be the element of $\mathscr{F}$ consisting of all indices $\geqslant(F^*, k, p)$. For each index $i\in V$, the mapping taking each a onto $f_a(i)$, $a\in F$, and each b onto $f_b(i)$, $b\in D'$, is a (k, p)-isomorphism of R* onto R. For each $i\in U$, the mapping taking $f_a(i_0)$ onto $f_a(i)$, $a\in F$, and $g(i_0)$ onto $g(i)$, $g\in D$, is a (k, p)-automorphism of R. The composition of this mapping with t_0^* yields a (k, p)-isomorphism of R* onto R which maps each a onto $f_a(i)$, $a\in F$, and each b onto $g(i)$, $b\in D'$. Finally, for each $i\in U\cap V$, the mapping fixing $f_a(i)$, $a\in F$, and taking each $g(i)$ onto the corresponding $f_b(i)$, $g\in D$, is a (k, p)-automorphism of R.

The proof of part 1 is a simplification of that of part 2: here R* is assumed to be only logically equivalent to R and we do not assume that the functions f_a, $a\in E$, are constant. ◁

7.5.7. By a theorem of [FRAY-MOR-SCO, 1962], p. 217, *if* R *and* S *are logically equivalent multirelations, there exists a maximal ultrapower of* R *which is isomorphic to a logical extension of* S. This theorem is another version of 7.5.6 (1) (use 7.5.4).

By a theorem of [KOC, 1961], p. 241, *if* R* *is a logical extension of* R, *there exist a maximal ultrapower* T *of* R *and an isomorphism* f *of* R* *onto a logical restriction* f(R*) *of* T *such that each element* a *of the base* |R| *is mapped onto the constant function with value* a. This is another version of 7.5.6(2).

EXERCISES

1

Let R be a multirelation with base E, I an index set and B the set of functions mapping I into E. With each $i \in \mathrm{I}$ we associate the mapping of B into E, again denoted by i, defined by $i(f)=f(i)$ for each $f \in \mathrm{B}$. Let $\mathscr{F}$ be an ultrafilter on I; by the above procedure, $\mathscr{F}$ may be viewed as an ultrafilter on the mappings of B into E.

(1) Given an n-tuple of elements $f_1, \ldots, f_n \in \mathrm{B}$, the value $\mathrm{S}(f_1, \ldots, f_n)$ of the maximal ultrapower S is + or − according as $\mathrm{R}(i(f_1), \ldots, i(f_n))$ is + or − for almost all mappings i. We thus get the inverse mapping of R under the projection ultrafilter defined on the mappings of B into E (see 4.3).

(2) Two functions f and g are equivalent mod $\mathscr{F}$ if and only if they are equivalent as elements of B in the sense of 4.3.2, in other words, if $i(f)=i(g)$ for almost every mapping i.

(3) A function f is constant, taking the fixed value $a \in \mathrm{E}$, if $i(f)=a$ for every mapping i. A constant function of this type may be identified with the element $a \in \mathrm{E}$. Show that the mappings i whose restriction to E is the identity mapping, i.e., the mappings taking each constant a onto $i(a)=a$, constitute an element of the ultrafilter.

This exercise was suggested by E. Engeler.

2

Let R be a multirelation with infinite base E, I the set of all finite subsets F of E. For each F, let $\mathrm{U_F}$ be the set of all finite subsets F′ such that $\mathrm{F} \subseteq \mathrm{F}' \subseteq \mathrm{E}$. Let $\mathscr{F}$ be an ultrafilter on I which is finer than the set of all the sets $\mathrm{U_F}$. With each F, associate the restriction $\mathrm{R} \mid \mathrm{F}$, and with each $a \in \mathrm{E}$ a function f_a such that $f_a(\mathrm{F})=a$ for $a \in \mathrm{F}$, $f_a(\mathrm{F})$ is arbitrary for $a \notin \mathrm{F}$. The result is an ultraproduct isomorphic to R. If this ultraproduct is normal, then R is discrete in the sense of Chapter 4, Exercise 1. Is the converse true?

3

Let I be a nondenumerable set. The cardinal of I is said to be *ω-measurable* if there exists a nontrivial ultrafilter $\mathscr{F}$ on I such that any denumerable intersection of elements of $\mathscr{F}$ is in $\mathscr{F}$. Show that the following definitions are equivalent: (a) For any denumerable partition of I, one of the classes is in $\mathscr{F}$. (b) For any denumerable relation R, the maximal ultrapower of R modulo $\mathscr{F}$ is isomorphic to R.

CHAPTER 8

FORCING

The method of forcing starts from a multirelation R and a natural number n and defines the properties of a certain n-ary relation S, known as a general relation, on the base $|R|$. The relation S is in a sense "in general position" relative to R. For example, if R is the chain of natural numbers and $n=1$, the unary general relation S is $+$ for infinitely many numbers and $-$ for infinitely many numbers; S is $+$ for infinitely many even numbers and $-$ for infinitely many even numbers, and so on. Perfected by Paul Cohen in 1963, forcing was the main tool in his proof of the independence of the axiom of choice and the generalized continuum hypothesis.

8.1. Generic predicate; system; (+)-forced and (−)-forced formulas

Let R be a multirelation with base E and n a natural number. We call the pair $\sigma=(R, n)$ an *n-ary generic predicate* or *generic predicate of arity n*. A *system assignable* to σ is any finite set of n-tuples of elements of E, each associated with a value $+$ or $-$.

An (R, σ)-*formula* is a logical formula whose predicates are replaced by relations of R and by σ (this implies the assumption that these predicates are replaceable by the relations of R and an n-ary relation), with elements of E substituted for the free indices.

Let U be a system and P an (R, σ)-formula. We shall say that U (+)-*forces* or (−)-*forces* P if the following conditions hold.

First let P be a free operator (with the predicates replaced by R, σ and the free indices by elements of E). Let S be any completion of the "partial relation" defined by the system U on E (S is an n-ary relation with base E, taking the value specified in U for n-tuples in the system and arbitrary values otherwise). There are three possibilities: (a) P always takes the value $+$, regardless of the specific completion chosen; then we say that U (+)-*forces* P. (b) P always takes the value $-$, regardless of the specific completion chosen; then we say that U $(-)$-*forces* P. (c) P may take either value $+$ or $-$; then U does not force P. Note that, essentially, S need be defined only for the finite set of n-tuples whose terms are the elements of E substituted for the indices of P.

Now let P have the form $\alpha\ P_1 \ldots P_h$, where α is an h-ary connection and $P_1, \ldots, P_h$ are formulas. Substitute $+$ for those of the formulas which are $(+)$-forced and $-$ for those which are $(-)$-forced. If α takes the value $+$ regardless of the values substituted for the non-forced formulas, we say that U $(+)$-*forces* P. If α takes the value $-$ regardless of the values substituted for the non-forced formulas, we say that U $(-)$-*forces* P. Otherwise, U does not force P.

If P has the form $\underset{i}{\exists} Q(x^i)$, where $Q(x^i)$ denotes a formula in which all free indices except i are replaced by elements of E, we associate with each $a \in E$ the (R, σ)-formula $Q(a)$ obtained from $Q(x^i)$ by substituting a for i. We say that U $(+)$-*forces* P if there exists an element $a \in E$ such that U$(+)$-forces $Q(a)$. U $(-)$-*forces* P if, for any $a \in E$ and any system $V \supseteq U$, the system V does not $(+)$-force $Q(a)$. If P has the form $\underset{i}{\forall} Q(x^i)$, we say that U $(+)$-*forces* P if, for any $a \in E$ and any system $V \supseteq U$, the system V does not $(-)$-force $Q(a)$. U $(-)$-forces P if there exists $a \in E$ such that U $(-)$-forces $Q(a)$.

The above definitions may be generalized, replacing n by a finite sequence ν of natural numbers. The pair $\sigma = (R, \nu)$ will then be called a *ν-ary generic multipredicate*. A system *assignable* to σ is then a union of systems assignable to each generic predicate figuring in σ. If $\nu = (n_1, \ldots, n_h)$, the system consists of finitely many n_1-tuples, each associated with a value $+$ or $-, \ldots,$ and finitely many n_h-tuples, each associated with a value $+$ or $-$. The above definitions of (R, σ)-formula and forcing extend immediately to generic multipredicates.

[If a system U $(+)$-forces a formula P, we shall call $+$ the *forced value* of P; we shall sometimes say that U forces P to the value $+$. A similar convention will be adopted for $(-)$-forcing.]

8.1.1. Suppose that the predicates replaced by σ in each of the free operators figuring in P are inactive. It is then readily seen that the forced value exists and is the same for any assignable system: *it coincides with the truth value of* P *for* R *and the base elements substituted for the free indices* (as defined in Volume 1, 5.2).

8.1.2. EXAMPLE: THE CHAIN OF NATURAL NUMBERS WITH A UNARY GENERIC PREDICATE. Let $\leqslant$ denote the chain of natural numbers, and let σ be a unary generic predicate over the natural numbers. In this case, a system is a finite set of natural numbers, each associated with + or −. The system $U=\{0\}$, with $\sigma(0)=+$, (+)-forces the formula $\sigma 0$ and hence also the formula $\exists_x \sigma x$. The only active predicate in the formula $\forall_y 0\leqslant y$ is $\leqslant$; by 8.1.1, any system forces this formula to its truth value, which is +. Thus U (+)-forces the conjunction $\sigma 0 \wedge \forall_y 0\leqslant y$, hence also $\exists_x(\sigma x \wedge \forall_y x\leqslant y)$. Call this formula P.

Now let $U=\{0\}$, with $\sigma(0)=-$. Then U (−)-forces the formula P. Indeed, for any natural number a and any system $V \supseteq U$, either $a=0$ and V does not (+)-force σa, or $a\neq 0$ and V does not (+)-force $\forall_y a\leqslant y$.

We now show that any system forces the set of natural numbers for which σ is + to be infinite, and the same holds for the complement of this set. For example, consider the formula $\forall_x \exists_y x\leqslant y \wedge \sigma y$; we claim that it is always (+)-forced. For any a and any system U, the latter does not (−)-force $\exists_y a\leqslant y \wedge \sigma y$, for there exist a number b (any number greater than a and not contained in U) and a system $V \supseteq U$ with $\sigma(b)=+$; thus V (+)-forces the formula $a\leqslant b \wedge \sigma b$. One shows similarly that any system (+)-forces the formula $\forall_x \exists_y x\leqslant y \wedge \neg \sigma y$.

The same method shows that any system (+)-forces the formula

$$\forall_t \exists_x \exists_y t\leqslant x \wedge x\leqslant y \wedge x\not\equiv y \wedge \sigma x \wedge \sigma y \wedge \forall_z (z\leqslant x \vee y\leqslant z),$$

which states that there are infinitely many pairs of consecutive natural numbers giving σ the value +. The same holds if we negate either (or both) of the formulas σx, σy; this yields the existence of infinitely many pairs of consecutive natural numbers giving σ the values + +, + −, − +, or − −. In general: for any h and any preassigned h-tuple of truth values + and −, the formula stating that there exist infinitely many sequences of h consecutive numbers giving σ these truth values is always (+)-forced.

There are formulas which may be either (+)-forced or not forced, but never (−)-forced. For example, consider the formula $\exists_x \sigma x$ and let a be an arbitrary natural number. This formula is $(+)$-forced by any system for which $\sigma(a) = +$, and it is not forced by a system all of whose truth values are $-$.

8.2. Elementary properties

8.2.1. *No system can both* $(+)$-*force and* $(-)$-*force the same formula.*

$\rhd$ Let P be a formula and U a system. If P is a free operator, there is nothing to prove. If P is $\alpha P_1 \ldots P_h$, where α is an h-ary connection and the assertion is true for $P_1, \ldots, P_h$, it is clearly true for P. Now let P have the form $\underset{i}{\exists}(Q(x^i)$ and suppose the assertion true for $Q(a)$, for any element a and any system. If U $(+)$-forces P, there exists an element a such that U $(+)$-forces $Q(a)$. If U also $(-)$-forces P, then for any element, a in particular, and any system containing U, U in particular, the system does not $(+)$-force $Q(a)$. This is a contradiction. The reasoning for $\forall$ is analogous. $\lhd$

8.2.2. *If a system* U *forces* P, *then any system* $V \supseteq U$ *forces* P *to the same value.*

$\rhd$ If P is a free operator, there is nothing to prove. If P has the form $\alpha P_1 \ldots P_h$, where α is an h-ary connection and the assertion is true for $P_1, \ldots, P_h$, any system $V \supseteq U$ forces (at least) those of the formulas $P_1, \ldots, P_h$ which are forced by U, and to the same value; thus V forces P to the same value. Now let P have the form $\underset{i}{\exists} Q(x^i)$ and suppose the assertion true for $Q(a)$, where a is arbitrary. If U $(+)$-forces P, there exists a such that U $(+)$-forces $Q(a)$; by the induction hypothesis, any system $V \supseteq U$ $(+)$-forces $Q(a)$ and hence also P. If U $(-)$-forces P and $V \supseteq U$, then, for any a and any system containing U, particularly for any system containing V, the formula $Q(a)$ is not $(+)$-forced; thus V $(-)$-forces P. $\lhd$

8.2.3. *For any formula* P *and any system* U, *there exists a system* $V \supseteq U$ *such that* V *forces* P *to either* $+$ *or* $-$.

$\rhd$ Let R be the given multirelation, E its base, and σ the generic predicate. First let P be a free operator. Then we need only enlarge U, assigning arbitrary values to the generic predicate σ for all n-tuples of elements substituted for the free indices of P. Now let P have the form $\alpha P_1 \ldots P_h$, where α is an h-ary connection. By the induction hypothesis, there exist a system $V_1 \supseteq U$ which forces P_1 to a certain value, then a system $V_2 \supseteq V_1$ which forces P_2 to some value,..., and finally a system V_h forcing P_h – this is the required system V. If P has the form $\exists_i Q(x^i)$, there are two possibilities: (a) There exist a and $V \supseteq U$ such that V (+)-forces $Q(a)$; then V (+)-forces P. (b) For no a and no system $V \supseteq U$ is the formula $Q(a)$ (+)-forced; then U (−)-forces P. Similar reasoning disposes of the case $P = \forall_i Q(x^i)$. ◁

8.2.4. A system may force a formula P without necessarily forcing every formula equideducible from P. Consider the example of 8.1.2 (the chain of integers with a unary generic predicate σ). The system $\sigma(0) = +$ (+)-forces the thesis $\sigma 0 \vee \neg \sigma 0$. However, the formula $\forall_x \sigma x$ is neither (+)- nor (−)-forced; the same holds for $\exists_x \neg \sigma x$. Thus the thesis $\forall_x \sigma x \vee \exists_x \neg \sigma x$ is not forced. Another example: the thesis $\forall_x \sigma x \vee \exists_y \neg \sigma y$ is not forced, although our system $\sigma(0) = +$ (+)-forces the thesis $\forall_x (\sigma x \vee \exists_y \neg \sigma y)$, which is obtained from the preceding formula by simply permuting $\vee$ and $\forall_x$. Indeed for any a and any system V containing our system, V does not (−)-force $\sigma a \vee \exists_y \neg \sigma y$.

8.3. Forcing with constraints

In addition to the multirelation R and n-ary generic predicate σ, suppose we are given a (possibly infinite) set of n-tuples $a_1, \ldots, a_n$ of base elements, called *constrained* n-tuples, each associated with a truth value $\sigma(a_1, \ldots, a_n) = +$ or $-$. These equalities are known as *constraints*. We shall now consider systems, defined as before by a finite set of n-tuples and corresponding truth values, with the added stipulation that the value of the system for any constrained n-tuple must be that given by the constraint.

We can then define (R, σ)-formulas and forcing as in 8.1, except that all the systems figuring in the definition associate the constraint value with any constrained n-tuple. Propositions 8.2.1 and 8.2.3 remain valid with this modified forcing concept.

Two extreme cases are possible here. In the absence of constraints, we have the original definition of 8.1. If every n-tuple is constrained, we have in effect replaced the generic predicate σ by an n-ary relation S; the forced value no longer depends on the system and simply reproduces the usual truth value for the multirelation (R, S) and the elements substituted for the free indices.

8.3.1. Let us return to the chain of natural numbers with a unary generic predicate σ, adding constraints $\sigma(a)=+$ for every even number a. Then any system will $(+)$-force the following formula, which states that for any two consecutive elements a, b, either $\sigma(a)=-$ or $\sigma(b)=+$:

$$\underset{x}{\forall}\underset{y}{\forall}(x \leqslant y \bigwedge x \not\equiv y \bigwedge \underset{z}{\forall}(z \leqslant x \bigvee y \leqslant z)) \Rightarrow (\sigma x \bigvee \sigma y).$$

Indeed, if we delete the quantifiers and replace x by a and y by b, then either $b \neq a+1$ and the antecedent is $(-)$-forced, or $b=a+1$ and the consequent is $(+)$-forced thanks to the constraints.

Other, analogous examples can be given: any system $(+)$-forces the formula which states that there exist infinitely many numbers such that $\sigma(a)=+$, or infinitely many numbers such that $\sigma(a)=-$, infinitely many pairs (a, b) such that $b=a+2$, $\sigma(a)=+$ and $\sigma(b)=-$, etc.

8.3.2. Consider a binary relation, which we call *membership* and denote by $\in$, defined on *herefinite* sets (abbreviation of "hereditarily finite" sets). Herefinite sets are defined as follows. The empty set 0 is herefinite. If a and b are herefinite, the set $a \cup \{b\}$ (the elements of a plus the element b) is herefinite. Every herefinite set is obtained from the empty set in this way; more precisely, the set of herefinite sets is the intersection of all sets which contain 0 and, whenever they contain a and b, also contain $a \cup \{b\}$. Thus $\{0\}$, $\{\{0\}\}$, $\{0, \{0\}\}$ are herefinite sets. We shall say that a set a is *transitive* if $x \in y$ and $y \in a$ imply $x \in a$, for any x, y. A set a is said to be *totally ordered* (by membership) if, for any $x \in a$ and $y \in a$, we have either $x \in y$, $y \in x$ or $x=y$. Define a natural number to be a totally ordered transitive herefinite set. Examples: 0, the singleton $1=\{0\}$, the pair $2=\{0,1\}=\{0, \{0\}\}$. The order relation $a<b$ on natural numbers coincides with the membership relation $a \in b$.

Let E be the set of herefinite sets. We view this set as the base for the membership relation $\in$ and a unary generic predicate σ with constraints $\sigma(u)=-$ for any element u in E which is not a natural number. Any system $(+)$-forces the following formula, which states that σ takes the value + for infinitely many elements (all necessarily natural numbers):

$$\forall_x (\sigma x \Rightarrow \exists_y (x \in y \wedge \sigma y)).$$

Indeed, for any a in E, if a is not a natural number, then $\sigma(a) = -$ by virtue of the constraints, and so $\sigma a \Rightarrow \exists_y (a \in y \wedge \sigma y)$ is $(+)$-forced by any system; hence it is not $(-)$-forced. If a is a natural number, the only case to be considered is that in which $\sigma(a) = +$. Even in this case, the system, U say, does not $(-)$-force the above formula. Otherwise, it would have to $(-)$-force the formula $\exists_y (a \in y \wedge \sigma y)$. Hence, for any b and any system $V \supseteq U$, V does not $(+)$-force the formula $a \in b \wedge \sigma b$. Now we need only let b be a natural number greater than all numbers in U and define V by the condition $\sigma(b) = +$.

As another example, let us show that any system $(+)$-forces the following formula, which states that there is no set containing all the natural numbers which satisfy σ: $\forall_u \exists_x x \notin u \wedge \sigma x$. Let U be a given system; we need only show that, for any a and any $V \supseteq U$, V does not $(-)$-force the subformula $\exists_x x \notin a \wedge \sigma x$. In other words, for all a and V there exist an element b and a system $W \supseteq V$ such that W $(+)$-forces the formula $b \notin a \wedge \sigma b$. It is now sufficient to let b be a natural number $> a$ and not figuring in V; W will be the union of V and the condition $\sigma(b) = +$.

8.4. General relation

Given a multirelation R and an n-ary relation S with the same base, consider the equalities $S(a_1, \ldots, a_n) = +$ or $-$, where $a_1, \ldots, a_n$ are arbitrary elements of the base. Any finite set of such equalities constitutes a system in the sense of 8.1, which defines a forcing concept for the n-ary generic predicate (R, n). Suppose that for every logical formula P with predicates replaced by R and an n-ary generic predicate, and free indices replaced by base elements, there exists a system consisting of finitely many equalities $S(a_1, \ldots, a_n) = +$ or $-$ which forces P to the value + or $-$. Then we shall say that S is a *general relation* for R. Since any two systems determined by the values of S may be embedded in a common super-system, it follows from 8.2.1 and 8.2.2 that the *forced value of* P *is always the same*, provided the system is large enough to force P.

The definition of a general relation is the same for forcing with constraints. Namely, we are given a (generally infinite) set of *constrained n-tuples*, and the definition of forcing is restricted to systems whose value for each constrained n-tuple $a_1, \ldots, a_n$ is the value $S(a_1, \ldots, a_n)$.

For example, let R be the chain of natural numbers, and add a unary relation S on the same base, with no constraints. If S is always $+$, it is not a general relation. Indeed, consider a finite set of natural numbers $u, \ldots, v$; the equalities $S(u) = \cdots = S(v) = +$ do not force the formula $\underset{x}{\exists} \neg \sigma x$ either to $+$ or to $-$: if we take any number x not in the set $\{u, \ldots, v\}$ and put $S(x) = -$, the resulting system forces $\neg \sigma x$ to $+$.

8.4.1. If all n-tuples are constrained, S is a general relation, and for every formula P the forced value is identical with the truth value of P for the multirelation (R, S) and the elements substituted for the free indices.

8.4.2. *Let* R *be a denumerable multirelation, n a natural number, and* U *a finite system of n-tuples provided with values + or −. There exists at least one n-ary general relation for* R *which satisfies the equalities of* U. (This proposition remains valid for forcing with constraints.)

▷ Enumerate all formulas, say $P_1, \ldots, P_i, \ldots$ and let $U_1 \supseteq U$ be a system forcing P_1 to one of the values $+$ or $-$ (see 8.2.3). Now let $U_2 \supseteq U_1$ be a system forcing P_2 to a suitable value, and so on. We thus get a sequence U_i $(i = 1, 2, \ldots)$, whose union specifies a value $S(a_1, \ldots, a_n)$ for each n-tuple $a_1, \ldots, a_n$, since the formula $\sigma a_1 \ldots a_n$ is one of the terms in the sequence $\{P_i\}$. ◁

8.4.3. *Let* R *be a multirelation and* S *a general relation for* R. *For every formula* P, *the forced value is the truth value assumed by* P *for the multirelation* (R, S).

▷ The assertion is obvious if P is free. Now suppose the assertion true for formulas $P_1, \ldots, P_h$ and let α be an h-ary connection; consider the formula $\alpha P_1 \ldots P_h$. For each P_i $(i = 1, \ldots, h)$, there exists a system U_i, consisting of finitely many equalities satisfied by S, which forces P_i to a certain value $v_i = +$ or $-$. The system $\bigcup_{i=1}^{h} U_i$ forces P to the value $v = \alpha(v_1, \ldots, v_h)$. By the induction hypothesis, v_i is the truth value of P_i

for (R, S), $i=1,\ldots, h$: thus the truth value of P is v.

Now let $P=\underset{i}{\exists} Q(x^i)$, where the assertion is true for Q. If P is (+)-forced this means that there is a base element a such that $Q(a)$ is (+)-forced. By hypothesis, the truth value of $Q(a)$ is +, and so that of P is also +. If P is (−)-forced by some system U consisting of equalities for S, this means that for no element a in the base is there a system $V \supseteq U$ that forces $Q(a)$ to +. Since S is a general relation, it follows that for every a there is a system of equalities for S which forces $Q(a)$ to some value, which cannot be + and is therefore −. By hypothesis, the truth value of $Q(a)$ is −; since this is true for every a, the truth value of P is −. A similar argument completes the proof for $P=\underset{i}{\forall} Q(x^i)$. ◁

8.4.4. *There exist relations* R, S, R′, S′, *where* S *is a general relation for* R, *such that* R′S′ *is a logical extension of* RS *but* S′ *is not general for* R′ (communicated by M. Pouzet).

▷ Let R be the chain of natural numbers and S a unary general relation for R. Let R_i, S_i be copies of R, S, respectively, $i=1, 2, \ldots$. Let f be a function from natural numbers to natural numbers, such that (a) $S(f(i))=+$ for each i; (b) the number of consecutive integers $f(i)+1, f(i)+2, \ldots$ such that $S(f(i))=S(f(i)+1)=S(f(i)+2)=\cdots=+$ tends to infinity with i. Such a function exists, since any finite sequence of truth values + or − may be realized on a suitable sequence of consecutive integers (see 8.1.2). Let $\mathscr{F}$ be an ultrafilter on the natural numbers and R′S′ the corresponding maximal ultrapower, which is a logical extension of RS. The function f is an element of the base |R′S′|.

We claim that S′ is not a general relation for R′. There exist infinitely many elements consecutive to f (mod R′) on which S′ takes the value +. On the other hand, R′S′ is logically equivalent to RS, and so for every element f of |R′S′| there exists a smaller element g (mod R′) such that $S'(g)=-$. Now consider the formula

$$\underset{x}{\forall}(f \leqslant x \wedge x < g (\operatorname{mod} R')) \Rightarrow \sigma x;$$

this formula takes the value + when S is substituted for σ. Thus, if S′ is a general relation for R′, the formula is not forced to −. Suppose that it is (+)-forced for some system U which is a restriction of S′. For any element $h \in |R'S'|$ and any system $V \supseteq U$, V does not (−)-force the formula

$(f \leqslant h \wedge h < g) \Rightarrow \sigma h$. But this is false, for the interval from f to g (mod R′) is infinite, and so there exist an element h in this interval which is not in U and a system V ⊇ U, such that $\sigma(h) = -$. ◁

Problems. (1) If S is general for R and R′S′ is a logical restriction of RS, is S′ general for R′?

(2) If S is general for R, is the negation ¬S general for R (or for ¬R, which is equivalent by 8.4.5 below) (communicated by C. Gnanvo)?

8.4.5. *If a multirelation* R′ *is interpretable in* R, *then any general relation for* R *is also general for* R′.

▷ Suppose that R′ is interpretable in R via a formula P, i.e., R′ = P(R). Let σ be a generic predicate defined for either R or R′, i.e., the definition of a system U assignable to σ does not depend on whichever of R or R′ is chosen but only on the elements of their common base. If U forces an (R′, σ)-formula Q(R′, σ), then U forces (to the same value) the (R, σ)-formula Q(P(R), σ) obtained by substituting P(R) for R′. Conversely, if U forces the second of these formulas, it forces the first to the same value. ◁

Problem. Let R be the chain or succession relation on the natural numbers, and S a unary relation on the natural numbers such that, for any finite sequence of h truth values + or −, there exist h consecutive numbers giving S these values. Is S necessarily a general relation for R?

8.5. Forcing and Deduction; Theory Forced by a Generic Predicate

8.5.1. *A thesis is never* (−)*-forced.*

▷ Let P be a thesis whose predicates are replaced by the multirelation R and the generic predicate, and U a system which (−)-forces P. By 8.4.2, there exists a general relation S for R which satisfies the equalities of U. By 8.4.3, the forced value − is also the truth value assumed by P for (R, S) and the base elements substituted for its free indices. But this contradicts the assumption that P is a thesis. ◁

Recall that a thesis is not necessarily forced by a given system. Examples were given in 8.2.4.

8.5.2. *If a system* U (+)*-forces* P *and* Q *is deducible from* P, *then*:

(1) U *does not* (−)*-force* Q.

(2) *There exists a system* $V \supseteq U$ *which* $(+)$*-forces* Q.

This follows from 8.5.1, the fact that $P \Rightarrow Q$ is a thesis, and 8.2.3.

Two equideducible formulas cannot be forced by the same system to different values.

8.5.3. Given a multirelation R and a generic predicate (or multipredicate) σ, *the set of bound formulas which are not* $(-)$*-forced by any system is a theory*, which we shall call the theory *forced by* R *and* σ. That this set of formulas is indeed a theory follows from the fact that it is closed under deduction and conjunction (see 8.5.2).

The theory forced by R *and an n-ary generic predicate is precisely the theory of all multirelations* RS, *where* S *ranges over all n-ary general relations for* R.

We shall in fact prove a more general assertion. Given a multirelation R, a natural number n, and an n-ary system U, the set of bound formulas which are not $(-)$-forced by any system $V \supseteq U$ is closed under deduction and conjunction and therefore constitutes a theory.

The theory just defined is the theory of all multirelations RS, *where* S *ranges over all n-ary general relations for* R *which are extensions of* U.

$\triangleright$ Let σ be an n-ary generic predicate and P an (R, σ)-formula. If P is not $(-)$-forced by any extension of U then, for any general relation S which is an extension of U, there exists a system V $(U \subseteq V \subseteq S)$ which forces P, necessarily to the value $+$; thus $P(S) = +$. Conversely, if P is forced to $-$ by some extension $V \supseteq U$, then by 8.4.2 there is a general relation S which is an extension of V and so satisfies $P(S) = -$. $\triangleleft$

Problem. Let R and R′ be logically equivalent. Is the theory forced by R and an n-ary generic predicate the same as that forced by R′ and an n-ary generic predicate?

8.5.4. Let E be a denumerable set, viewed as a multirelation.

An n-ary relation S *is general for* E *if and only if* S *is homogeneous and every denumerable n-ary relation is embeddable in* S (communicated by M. Boffa).

Recall that S is said to be *homogeneous* if any local automorphism of S with finite domain is extendible to an automorphism of S (see Volume 1, Chapter 4, Exercise 5). T is embeddable in S if S has a restriction isomorphic to T (see Volume 1, 3.2.3).

▷ Any two homogeneous relations in which any other denumerable relation is embeddable are isomorphic (Volume 1, Chapter 4, Exercise 5.3). It will therefore suffice to show that if S is a general relation for E it satisfies the stipulated conditions. Essentially, we must prove the following assertion: for any p, q, any sequences of elements $a_1,\ldots,a_p$ in E and $b_1,\ldots,b_q$ not in E, and any extension T of S $|$ $\{a_1,\ldots,a_p\}$ to $\{a_1,\ldots,a_p, b_1,\ldots,b_q\}$, there exists an isomorphism of T onto a restriction of S which acts as the identity mapping on $a_1,\ldots,a_p$ (see Volume 1, Chapter 4, Exercise 5.1).

Each of the above conditions, depending on p, q, the restriction S $|$ $\{a_1,\ldots,a_p\}$ and T, is stated by a formula of the following type (where σ is a binary predicate replaceable by S):

$$\underset{1,\ldots,p}{\forall}\ \underset{p+1,\ldots,p+q}{\exists}\ (x^1 \not\equiv x^2 \wedge \cdots \wedge \sigma x^1 x^1 \wedge \neg \sigma x^1 x^2 \wedge \cdots) \Rightarrow$$
$$\Rightarrow (x^{p+1} \not\equiv x^1 \wedge \cdots \wedge \sigma x^1 x^{p+1} \wedge \neg \sigma x^{p+1} x^{p+1} \wedge \cdots),$$

where the antecedent of the implication describes S $|$ $\{a_1,\ldots,a_p\}$ and the consequent its extension T. Since S is general for E, the truth value of a formula for S coincides with the value forced by a certain finite restriction of S. To prove that the above formula is + for S, it will suffice to show that it is not $(-)$-forced by any finite restriction of S. Indeed, since S is general, this would imply the existence of a system which is a restriction of S and forces the formula, necessarily to +.

We shall prove, in fact, that for *any* relation S with base E, no finite restriction U of S forces our formula to $-$. An equivalent assertion is that, for any elements $a_1,\ldots,a_p \in$ E, U does not $(-)$-force the formula

$$\underset{p+1,\ldots,p+q}{\exists}\ (a_1 \not\equiv a_2 \wedge \cdots \wedge \sigma a_1 a_1 \wedge \cdots) \Rightarrow$$
$$\Rightarrow (x^{p+1} \not\equiv a_1 \wedge \cdots \wedge \sigma a_1 x^{p+1} \wedge \cdots).$$

This, in turn, is equivalent to the statement that for any $a_1,\ldots,a_p \in$ E there exist elements $b_{p+1},\ldots,b_{p+q}$ and a system V $\supseteq$ U such that V $(+)$-forces (and therefore gives the value + to) the free formula

$$(a_1 \not\equiv a_2 \wedge \cdots \wedge \sigma a_1 a_1 \wedge \cdots) \Rightarrow (b_{p+1} \not\equiv a_1 \wedge \cdots \wedge \sigma a_1 b_{p+1} \wedge \cdots).$$

Now this is true for any relation S. Indeed, if the restriction S $|$ $\{a_1,\ldots,a_p\}$ is not described by the antecedent of this formula, the latter certainly takes the value $+$. If the restriction is so described, we need only select elements $b_{p+1},\ldots, b_{p+q}$ distinct from $a_1,\ldots, a_p$ and from the elements of E occurring in U, and then to take a system $V \supseteq U$ which satisfies the consequent; this is always possible by suitable assignment of values to formulas of the type $\sigma a_1 b_{p+1}$. $\lhd$

Since any two homogeneous denumerable relations in which every finite relation is embeddable are isomorphic, the theory forced by E and a generic predicate is saturated.

In fact, a more general proposition is true:

Let R *be a multirelation with denumerable base* E *such that for any finite subset* F *of* E *there exists an automorphism* f *of* R *for which* f(F) *is disjoint from* F. *Then the theory forced by* R *is saturated* (communicated by A. M. Sette).

$\rhd$ Let P be a bound formula with R and the generic predicates substituted for the predicates of P. If some system U forces P to $-$, while another system V forces $\neg$P to $-$, then there exists an automorphism f of R such that f(V) contains only elements disjoint from those of U. Since f is an automorphism, the bound formula $\neg$P is ($-$)-forced by f(V) as well. Thus the union $U \cup f(V)$ ($-$)-forces both P and $\neg$P; hence it forces P to both $+$ and $-$, and this is a contradiction. $\lhd$

For example, if R is the chain of integers the forced theory is saturated for generic predicates of arbitrary arity.

Problem. Does there exist a multirelation R, on a nondenumerable base, for which there is no unary general relation? The same question may be asked for other arities.

1

Let C be the succession relation on the natural numbers. It can be shown as in 8.4 that the unary relation which is + for all numbers is not general for C. Show that the relation S taking the value + for the even numbers and − for the odd numbers is not general for C. To be precise: no subsystem of S forces the following true formula: $\forall_{xy}(Cxy \wedge Sx) \Rightarrow \neg Sy$.

Even more: no relation which is + for all even numbers can be general for C. Extend the result to multiples of 3, 4, ...

2

Let R be the sum of natural numbers and S a unary relation on the natural numbers such that $S(x) = +$ implies $S(2x) = +$ for any x. Show that S is not a general relation for R: no subsystem of S forces the formula

$$\forall_{xy}(Rxxy \wedge Sx) \Rightarrow Sy.$$

Show that there exists a relation S satisfying the above condition which, in addition, yields any given finite sequence of truth values + and − for some sequence of consecutive numbers. Hence deduce that if Problem 8.4.5 has a positive solution, then S is a general relation for the chain of integers but not for their sum.

3

Generalize the concept of constraint (8.3), defining a constraint to be any (R, σ)-formula A for which there exists S such that $A(R, S) = +$. Example: Let R be the unary relation defining the base E, and A a formula stating that "σ is a chain". Systems are again assumed to be compatible with constraints. We shall say that a system U (+)-forces a free formula P if any relation S on E that satisfies the constraint (hence, any chain S on E), which is an extension of E, is such that $P(R, S) = +$. The definition of (−)-forcing is analogous. The other parts of the definition (see 8.1) are unchanged.

Show that 8.2 and 8.4 remain valid.

Show that the general chain on E (i.e., the general relation for R = E with the constraint 'σ is a chain') is the dense chain with neither minimal nor maximal element.

CHAPTER 9

ISOMORPHISMS AND EQUIVALENCES IN RELATION TO THE CALCULUS OF INFINITELY LONG FORMULAS WITH FINITE QUANTIFIERS

9.1. α-ISOMORPHISM AND α-EQUIVALENCE

Let R, R′ be two multirelations with bases E, E′, respectively, f a local isomorphism of R onto R′, defined on $F \subseteq E$ with range $F' \subseteq E'$. We shall associate with f certain ordinals α, defined inductively as follows.

Every local isomorphism is a 0-*isomorphism*.

If $\alpha \geqslant 1$ and α has an immediate predecessor $\alpha - 1$, the mapping f is an α-*isomorphism* if, for any set $\bar{F}$ obtained by adjoining to F finitely many elements of E, there exists an extension $\bar{f}$ of f to $\bar{F}$ which is an $(\alpha - 1)$-isomorphism of R onto R′; and conversely with F, E, f, R, R′ replaced respectively by F′, E′, f^{-1}, R′, R.

If α is a limit ordinal, the mapping f is an α-*isomorphism* if it is a β-isomorphism for every $\beta < \alpha$.

A bijective mapping f of a subset of E onto another subset is an α-*automorphism* of R if it is an α-isomorphism of R onto R.

Every α-isomorphism is also an α'-isomorphism for all $\alpha' \leqslant \alpha$.

If a mapping f defined on F is an α-isomorphism of R onto R′, the same is true of the restriction of f to any subset of F.

An isomorphism of R onto R′ (mapping the base $|R|$ onto $|R'|$) is an α-isomorphism for any ordinal α. Thus any local isomorphism of R onto R′ which is extendible to an isomorphism of R onto R′ is an α-isomorphism for any ordinal α. For any subset F of $|R|$, the identity mapping on F is an α-automorphism of R for any ordinal α.

If f is an α-isomorphism of R onto R′, then f^{-1} is an α-isomorphism of R′ onto R. If moreover g is an α-isomorphism of R′ onto R″, then gf is an α-isomorphism of R onto R″.

9.1.1. A multirelation R′ is said to be α-*equivalent* to a multirelation R if the empty mapping is an α-isomorphism of R onto R′. It follows from the previous remarks that this relation is reflexive, symmetric and transitive. A sufficient (and clearly also necessary) condition for R and R′ to be α-equivalent is that there exist an α-isomorphism of R onto R′.

R *and* R′ *are* 1-*equivalent if and only if any restriction of* R *to a finite subset of its base is isomorphic to a restriction of* R′, *and conversely.*

It follows that, for any multirelation R and any infinite subset D of its base, there exists (assuming the axiom of choice) a superset D^+ of D, equipollent to D, such that the restriction $R \mid D^+$ is 1-equivalent to R. In fact, D^+ is obtained by adding to D all elements of a denumerable set of finite subsets F of $|R|$ such that any finite restriction of R is isomorphic to a restriction $R \mid F$.

9.1.2. In contradistinction to the situation for 1-equivalence, one cannot state in general that any multirelation R has a denumerable restriction which is 2-equivalent to R. We shall give an example of a birelation R of arity (1, 2), with base of the cardinal of the continuum, such that *the cardinal of any birelation* 2-*equivalent to* R *is at least that of the continuum.*

With each set A of natural numbers, we associate a binary succession relation C_A on the natural numbers: $C_A(x, y) = +$ if $y = x+1$, $C_A(x, x) = +$ if $x \in A$, $C_A(x, x) = -$ otherwise. We shall assume that the bases of the relations C_A for distinct A are disjoint; thus the union of the bases has the cardinal of the continuum. With this union as base, we define a birelation (Z, C): Z is the unary relation taking the value + for the zeros of each C_A, − otherwise; C is a binary relation, a common extension of the relations C_A, taking the value − for pairs (x, y) such that x and y lie in bases of different relations C_A.

▷ Let $R' = (Z', C')$ be a birelation which is 2-equivalent to R. For any given element a which is the zero of some C_A, corresponding to an element a' of the base of R′, the pair (a, a') constitutes a 1-isomorphism of R onto R′. To any finite sequence of consecutive elements beginning with a corresponds a sequence with the same number of terms from $|R'|$, and the mapping taking the first sequence onto the second is a local isomorphism of R onto R′. Hence it follows that C′ has a restriction isomorphic to the succession relation C_A whose zero is a. Any two of these restrictions necessarily have disjoint bases, for otherwise some element in the base of C′ would have two C′-successors (or two C′-predecessors), contrary to the 1-equivalence. Letting the set A vary, we see that the base of R′ has cardinal at least that of the continuum. ◁

9.1.3. Let R, R′ be two chains with bases E, E′, and f a local isomorphism of R onto R′ with domain F and range $F' = f(F)$. For the next proposition the reader should refer to the definitions of F-interval and the corresponding F′-interval in 1.2.7.

A local isomorphism is an α-isomorphism of R *onto* R′ *if and only if any two corresponding intervals are α-equivalent.* The proof is the same as in 1.2.7.

9.1.4. Let I be a chain. With each element i of the base of I we associate two chains R_i and R'_i with disjoint bases. Let R be the sum of the chains R_i as ordered by I, and R′ the sum of the R'_i similarly ordered. For each i, let f_i be a local isomorphism (possibly empty) of R_i onto R'_i, and let f be the union of the mappings f_i; f is a local isomorphism of R onto R′.

If f_i is an α-isomorphism of R_i onto R'_i for each i, then f is an α-isomorphism of R *onto* R′. *In particular, if R_i and R'_i are α-equivalent for each i, then* R *and* R′ *are α-equivalent.* This proposition is analogous to Exercise 2 in Chapter 1.

9.1.5. *Let α be an ordinal and* U, U′ *two chains (which may be empty). Then the two chains $\omega^\alpha + U(\omega^\alpha)$ and $\omega^\alpha + U'(\omega^\alpha)$ are α-equivalent.*

▷ The assertion is true for $\alpha = 0$, as any two binary relations are 0-equivalent. Suppose the assertion true for α and consider the chains

$$A = \omega^{\alpha+1} + U(\omega^{\alpha+1}) \quad \text{and} \quad A' = \omega^{\alpha+1} + U'(\omega^{\alpha+1}).$$

Let s be a finite sequence of elements $a \in |A|$; s defines a finite sequence of intervals of A, each of which is either an ordinal $r < \omega^\alpha$, or ω^α plus a sum of ordinals ω^α plus some $r < \omega^\alpha$. The last interval must be ω^α plus a sum of ordinals ω^α. We consider corresponding intervals in A′, as the case may be: an ordinal identical to r, or the sum $\omega^\alpha + r$. The elements a' corresponding to the elements a thus lie in the initial interval $\omega^{\alpha+1}$ of A′, and the last interval in A′ is ω^α plus a sum of ordinals ω^α. It follows that two corresponding intervals are either isomorphic or have the form $\omega^\alpha + U(\omega^\alpha) + r$ and $\omega^\alpha + U'(\omega^\alpha) + r$, where $r < \omega^\alpha$. By the induction hypothesis and 9.1.4 above (if necessary), any two corresponding intervals are α-equivalent. By 9.1.3, the mapping taking each a onto a' is an α-isomorphism. Interchanging A and A′, we see that these chains are $(\alpha+1)$-

equivalent. Finally, if α is a limit ordinal and the assertion is true for any ordinal $<\alpha$, it is clearly true for α. ◁

Let U be empty and $U' = \omega^-$ (the chain ω reversed); then A is an ordinal and A′ is not. Thus: *for any* α, *there exist* α-*equivalent chains one of which is an ordinal and the other is not* (see [KARP, 1965]).

9.1.6. *Let* α *be an ordinal,* $\alpha \geqslant 1$, *and* $A = \omega^\alpha$. *Then no ordinal* $<\omega^\alpha$ *is* α-*equivalent to* A.

▷ The assertion is trivially true for $\alpha = 1$. Suppose it true for α, let $A = \omega^{\alpha+1}$ and let $A' < A$ be an ordinal. Consider a finite sequence of elements a in the base of A, which determine initial intervals of type $i\omega^\alpha$ ($i = 0, 1, 2, \ldots$) the first of which is $\geqslant A'$. Suppose that A′ is $(\alpha+1)$-equivalent to A. Corresponding to these elements a we then have an equal number of elements a' in the base of A′, such that the mapping taking each a onto a' is an α-isomorphism. By 9.1.3, any two corresponding intervals are α-equivalent. Thus we have an interval of length at least ω^α which is α-equivalent to a strictly smaller interval – contradiction.

Now let α be a limit ordinal. Suppose the assertion true for any ordinal $<\alpha$; let $A = \omega^\alpha$ and $A' < A$. There exists an ordinal $\beta < \alpha$ such that $\omega^\beta > A'$. Let b be the element in the base of A which determines the initial interval ω^β. Then there exists an element b' such that the mapping taking b onto b' is a $(\beta+1)$-isomorphism of A onto A′. By 9.1.3, the initial intervals determined by b and b' are $(\beta+1)$-equivalent; thus ω^β is $(\beta+1)$-equivalent to a strictly smaller ordinal – contradiction. ◁

Consequently, *the ordinals* ω^α *and* $2\omega^\alpha$ ($\alpha \geqslant 1$) *are* α-*equivalent* (9.1.5) *but not* $(\alpha+1)$-*equivalent* (9.1.6).

Problem. Let R and R′ be α-equivalent multirelations. Do there exist extensions S, S′, α-equivalent to R, R′, respectively, which are isomorphic? The answer is positive for $\alpha = 1$ (see Volume 1, 3.2.6). The same question may be asked for the concept of ∞-equivalence defined in 9.2 below. A weaker version of the problem is as follows: Is it true that for each α there exists $\beta \geqslant \alpha$ such that, if R and R′ are β-equivalent, they have isomorphic extensions S and S′, α-equivalent to R and R′, respectively?

9.2. ∞-ISOMORPHISM AND ∞-EQUIVALENCE; KARPIAN FAMILIES

A local isomorphism of a multirelation R onto R′ is called an ∞-*isomorphism* if it is an α-isomorphism of R onto R′ for every ordinal α. A local automorphism of R is an ∞-*automorphism* if it is an α-automorphism of R for every ordinal α.

∞-isomorphism is preserved under restriction to any subset of its domain, under passage to the inverse mapping and under composition. An isomorphism of R onto R′ (or of any of its restrictions) is an ∞-isomorphism.

Multirelations R and R′ are said to be ∞-*equivalent* if they are α-equivalent for every ordinal α, or, equivalently, if the empty mapping is an ∞-isomorphism of R onto R′, or if there exists an ∞-isomorphism of R onto R′.

9.2.1. *Let f be an ∞-isomorphism of* R *onto* R′. *For any element a in the base of* R, *there exists an element a' in the base of* R′ *such that the union of f and the pair (a, a') is an ∞-isomorphism of* R *onto* R′.

$\triangleright$ For each α, the mapping f is an $(\alpha+1)$-isomorphism of R onto R′. Thus for any a there exists a' such that the union of f and the pair (a, a') is an α-isomorphism. Since the number of possible choices for the element a' is bounded by the cardinal of the base of R′, there exists an element a' satisfying the requirements for arbitrarily large ordinals, and so for every ordinal. $\triangleleft$

9.2.2. *Different ordinals cannot be ∞-equivalent.*

$\triangleright$ Let A be the smallest ordinal for which there exists an ∞-equivalent ordinal $A' > A$. Let a' be the element in the base of A′ defining the initial interval isomorphic to A. There exists an element $a \in |A|$ such that the pair (a, a') is an ∞-isomorphism. By 9.1.3, the corresponding initial intervals are ∞-equivalent, but one of them is smaller than A – contradiction. $\triangleleft$

9.2.3. Given two multirelations R and R′, a family I of local isomorphisms of R onto R′ is known as a *Karpian family* if it is nonempty and satisfies the following condition [KARP, 1965]:

For any mapping $f \in I$ and any element $a \in |R|$, there exists an element $a' \in |R'|$ such that the union of f and the pair (a, a') is in I; and conversely, with R and R′ interchanged.

Examples. By 9.2.1, the ∞-isomorphisms of R onto R′ (if such exist) form a Karpian family; the same is true of the ∞-isomorphisms with finite domains.

Let $R' = R$; then the identity mappings on all subsets of the base, or on all finite subsets, form a Karpian family.

Let R be the chain of rational numbers and R′ the chain of real numbers (or any other dense chain with neither minimal nor maximal element). Then the family of local isomorphisms of R onto R′ with finite domains is Karpian. This is not true for the family of all local isomorphisms: let R′ be an extension of R and consider the identity mapping defined on all of R.

9.2.4. For given R and R′, any union of Karpian families of local isomorphisms of R onto R′ is a Karpian family. Thus there exists a *maximal Karpian family*, viz. the union of all Karpian families. For example, if $R' = R$ has a denumerable base, we shall see in 9.4.1 below that this maximal family consists of the automorphisms of R and all their restrictions.

For given R *and* R′, *the maximal Karpian family is the set of all ∞-isomorphisms of* R *onto* R′.

R *and* R′ *are ∞-equivalent if and only if there exists a Karpian family of local isomorphisms of* R *onto* R′.

▷ By 9.2.1, the ∞-isomorphisms form a Karpian family. Conversely, we shall prove that any local isomorphism of R onto R′ in a Karpian family is an ∞-isomorphism. It is clearly an 0-isomorphism. Now, for any ordinal α, if every local isomorphism in a Karpian family is an α-isomorphism, it follows from the definition of Karpian families that every local isomorphism in the family is an $(\alpha+1)$-isomorphism. Finally, let α be a limit ordinal and suppose that f is a β-isomorphism for every $\beta < \alpha$. Then by definition it is also an α-isomorphism.

Now suppose that R and R′ are ∞-equivalent. Then the family of ∞-isomorphisms is not void and, by 9.2.1, it is Karpian. Conversely, suppose that there exists a Karpian family of local isomorphisms of R onto R′. By what we have just proved, these local isomorphisms are ∞-isomorphisms, and so R and R′ are ∞-equivalent. ◁

The examples cited in 9.2.3 show that *any two dense chains with neither maximal nor minimal element are ∞-equivalent* (e.g., the chain of rationals and the chain of reals).

9.2.5. In general one cannot define a minimal Karpian family, since there may be a situation in which any Karpian family has a proper Karpian subfamily. For example, if $R'=R$ the identity mappings on finite subsets of the base form a Karpian family, but the same is true of the identity mappings on sets containing at least p elements (for any natural number p). However, the family in question is minimal with respect to the Karpian families *closed under restriction*; a family I is closed under restriction if, whenever f is a member of I, the restriction of f to an arbitrary subset of its domain is also a member of I.

There may exist an infinite increasing sequence of Karpian families, all closed under restriction. Let R be the chain of rationals and R′ the chain of reals. For any real number r, the local isomorphisms of R onto R′ with finite domains such that the distance between any element of the domain and its image is always $<r$, form a Karpian family which is closed under restrictions. Letting r decrease, we obtain an increasing sequence as required.

Problems. If R and R′ are ∞-equivalent, does there exist a minimal Karpian family closed under restriction? Stronger version: does every Karpian family which is closed under restriction contain a minimal family with this property? Is this true, at least, for the chains of rational and real numbers?

9.3. Automorphic rank of a multirelation

Let R *be a multirelation. There exists an ordinal α such that the following equivalent conditions are satisfied for all $\beta \geqslant \alpha$:*

(1) *every β-automorphism of* R *with finite base is an ∞-automorphism;*

(2) *every β-automorphism of* R *with finite base is a $(\beta+1)$-automorphism;*

(3) *the β-automorphisms of* R *with finite bases form a Karpian family.*

Moreover, the cardinal of α is at most equal to the cardinal of $|R|$.

The ordinal α thus defined is known as the *automorphic rank* of R. For example, a dense chain with neither minimal nor maximal element is of rank 0, since its local automorphisms with finite domains form a Karpian family. A multirelation with finite base is of rank 0 or 1, since any 1-automorphism is extendible to an automorphism of the whole multirelation.

▷ Given an ordinal γ, let us say that two finite sequences $\{a\}$ and

$\{a'\}$, with the same number of terms in the base of R, are γ-equivalent if the mapping taking each a onto the corresponding a' is a γ-automorphism of R. For any pair of finite sequences, either there exists an ordinal γ such that the sequences are γ-equivalent but not $(\gamma+1)$-equivalent, or else the sequences are ∞-equivalent. Let α be the minimal ordinal strictly greater than all possible ordinals γ (for all possible pairs of finite sequences). Then α satisfies condition (1), hence also condition (2). It also satisfies condition (3), for the ∞-automorphisms with finite bases form a Karpian family. Let α be an ordinal satisfying (2). Then the family of α-automorphisms with finite domains is a Karpian family and therefore consists of ∞-automorphisms, so that α satisfies (1) and (3). Finally, if α satisfies (3), it follows from 9.2.4 that the α-automorphisms with finite bases are ∞-automorphisms, so that α satisfies (1) and (2).

It remains to prove that the cardinal of α is at most that of the base or, equivalently, at most that of the set of pairs of finite sequences of base elements. Let β be an ordinal, $\beta<\alpha$; there exists at least one pair of finite sequences which are β-equivalent but not $(\beta+1)$-equivalent, for otherwise α would not be the minimal ordinal satisfying (2). Since each pair of finite sequences determines a unique ordinal β in the above manner, the set of ordinals $<\alpha$ is at most equipollent to the base of R. ◁

We know of no interesting characterization of multirelations with finite or denumerable ranks.

9.3.1. *Let* A *be an ordinal; then the automorphic rank of* A *is at most* A.

▷ It suffices to show that any A-automorphism of A is the identity mapping. Supposing the contrary, let f be an A-automorphism mapping an element a_0 of the base onto some $a_1<a_0$ (mod A). Let α_0 and α_1 denote the ordinals of the initial intervals of A defined by a_0 and a_1. The pair (a_0, a_1) is an A-automorphism, hence also an α_0-automorphism of A. Hence there exists an element $a_2<a_1$ (mod A) such that the set consisting of the two pairs (a_0, a_1) and (a_1, a_2) constitutes an α_1-automorphism of A. Iterating this procedure, we see that there is an integer h such that a_h is the minimal element of A and the set of pairs (a_i, a_{i+1}) $(i=0, 1, \ldots, h-1)$ constitutes an α_{h-1}-automorphism of A (where α_i denotes the ordinal of the initial interval defined by a_i). Since $\alpha_{h-1}>0$ the element corresponding to a_h must precede it – contradiction. ◁

It is clear that the ordinals 0 and 1 are of rank 0; the finite ordinals

2, 3,... and the ordinal ω are of rank 1. We know of no interesting characterization of the ordinals of finite or denumerable rank.

9.3.2. *Let* R *and* R′ *be two multirelations of automorphic ranks* $\leqslant\alpha$. *If* R′ *is* $(\alpha+1)$*-equivalent to* R, *then* R *and* R′ *are* ∞*-equivalent and all* α*-isomorphisms of* R *onto* R′ *with finite domains are* ∞*-isomorphisms.*

▷ Let f be an α-automorphism of R onto R′ with finite domain F and range $F'=f(F)$, and let a be an element of the base of R. There exists an α-isomorphism h of R onto R′ whose domain is $F\cup\{a\}$. Set $b=h(a)$. The mapping fh^{-1} is an α-automorphism of R′, and hence an ∞-automorphism of R′, with domain $h(F)$. Thus there exists an element a' of the base of R′ such that $fh^{-1}\cup\{(b,a')\}$ is an ∞-automorphism of R′. Finally, since $f\cup\{(a,a')\}$ is the composite of h and the above ∞-automorphism, it is an α-isomorphism of R onto R′. The same arguments hold with R and R′ interchanged. Hence the family of α-isomorphisms of R onto R′ with finite domains is Karpian. Thus they are all ∞-isomorphisms, so that R and R′ are ∞-equivalent. ◁

The proposition fails to hold if R and R′ are only α-equivalent. For example, the ordinal ω and the reverse chain ω^- are both of rank 1, they have the same finite restrictions (up to isomorphism) and are therefore 1-equivalent, but they are not 2-equivalent.

9.3.3. *Let* R *be a multirelation of automorphic rank* α. *If* R′ *is* $(\alpha+2)$*-equivalent to* R, *then the rank of* R′ *is* α.

▷ Let h' be an α-automorphism of R′ with finite domain H′. There exists an $(\alpha+1)$-isomorphism f of R onto R′ whose range is $H'\cup h'(H')$. Set $H=f^{-1}(H')$ and $h=f^{-1}h'f$; the latter mapping, defined on H, is an α-automorphism of R and hence an ∞-automorphism. Thus the mapping $h'=fhf^{-1}$ is an $(\alpha+1)$-automorphism of R′. By 9.3, condition (2), the automorphic rank of R′ is $\leqslant\alpha$. Interchanging R and R′, we see that the rank is exactly α. ◁

9.3.4. *Let* R *be a multirelation and* ω_α *the cardinal of its base. There exists an ordinal* $\beta<\omega_{\alpha+1}$ *such that any multirelation* β*-equivalent to* R *is* ∞*-equivalent to* R.

▷ The automorphic rank γ of R has cardinal at most ω_α, by 9.3. Setting $\beta=\gamma+2$, we see that any multirelation β-equivalent to R is of rank γ (see 9.3.3), and hence ∞-equivalent to R by 9.3.2. ◁

9.3.5. Proposition 9.3.3 is false in general if R′ is only $(\alpha+1)$-equivalent to R. To prove this, we let N, Z, Q denote the chains of natural numbers, integers and rational numbers, respectively.

The chain Q(Z) *is of rank* 1.

▷ For any two finite sequences of pairs (a, b) and (a', b'), where a, a' are rational numbers and b, b' integers, the mapping taking each pair (a, b) onto a corresponding pair (a', b') and preserving lexicographic order is a 1-isomorphism if and only if any two corresponding intervals (see 1.2.7) are either both of the same finite cardinality or both infinite. Thus this mapping may be extended to an automorphism of Q(Z), and so it is an ∞-automorphism. ◁

The chain N(Z) *is of rank* $\geqslant 2$.

▷ Let f be the mapping taking the pair $(1, 0)$ onto $(0, 0)$; this mapping is a 1-automorphism since the corresponding intervals are infinite chains and are therefore 1-equivalent. It is not a 2-automorphism, for if we could extend f to the pair $(0, 0)$, the image of the latter would have to be a pair $(0, u)$, where u is a negative integer; we would thus have two corresponding intervals, one finite and the other infinite. ◁

The chains N(Z) *and* Q(Z) *are 2-equivalent.*

▷ Let us map each finite sequence of elements of the base of N(Z) onto itself (N being a restriction of Q in the usual sense of integers and rational numbers). Any two corresponding intervals are either of the same finite cardinality or both infinite. Thus the identity mapping in question is a 1-isomorphism. Conversely, consider an increasing sequence of pairs (a_i, b_i) $(i = 1, 2, \ldots, n)$, where the a_i are rationals and the b_i integers. With these pairs we associate the pairs (c_i, b_i), where $c_1 = 1$, $c_{i+1} = c_i$ if $a_{i+1} = a_i$, and $c_{i+1} = c_i + 1$ if $a_{i+1} \neq a_i$. Any two corresponding intervals are either of the same finite cardinality or both infinite, so that the mapping just defined is a 1-isomorphism (this example is due to J. P. Calais). ◁

9.4. Multirelations with Denumerable Bases and α-Isomorphisms

9.4.1. *Let* R, R′ *be two multirelations with denumerable bases. If* R *and* R′ *are* ∞*-equivalent, they are isomorphic. Moreover, any* ∞*-isomorphism of* R *onto* R′ *is extendible to an isomorphism of* R *onto* R′.

▷ Arrange the elements of the base of R in a sequence a_i ($i=0, 1, 2, \ldots$), and similarly for the elements of $|R'|$, a'_i ($i=0, 1, 2, \ldots$). Let f be an ∞-isomorphism of R onto R′ and let i be the smallest integer such that a_i is not in the domain of f. There exists an integer j such that $f \cup \{(a_i, a'_j)\}$ is an ∞-isomorphism. Repeating this procedure, alternating between the a with the smallest index not in the domain and the a' with the smallest index not in the range, we obtain a sequence of local isomorphisms, beginning with f, each an extension of its predecessor, such that the domains and ranges ultimately exhaust all the elements of the bases of R and R′. The union of these local isomorphisms is an isomorphism of R onto R′. ◁

9.4.2. *Let* R *be a multirelation with denumerable base whose automorphic rank* α *is a denumerable ordinal (see* 9.3*). Then:*

(a) *every* α*-automorphism of* R *with finite domain is extendible to an automorphism of* R*;*

(b) *any multirelation* R′ *which is* ($\alpha+2$)*-equivalent to* R *is* ∞*-equivalent to* R*; if the base of* R′ *is denumerable,* R′ *is isomorphic to* R.

▷ Every α-automorphism of R with finite domain is an ∞-automorphism, by 9.3; thus it is extendible to an automorphism of R, by 9.4.1.

Every multirelation R′ which is ($\alpha+2$)-equivalent to R is ∞-equivalent to R, by 9.3.2 and 9.3.3. If R′ is denumerable, it is isomorphic to R, by 9.4.1. ◁

Recall that even if R is denumerable there may be a nondenumerable multirelation R′ which is ∞-equivalent to R. An example is the chain R of rationals and the chain R′ of reals (see 9.2.4).

9.4.3. *Let* R *be a multirelation with denumerable base and* R′ *any multirelation. Then the* ω_1*-isomorphisms of* R *onto* R′ *(if such exist) are* ∞*-isomorphisms.*

If R *has a denumerable base, any* ω_1*-automorphism of* R *is extendible to an automorphism of* R.

▷ We shall prove that the ω_1-isomorphisms of R onto R′ constitute a Karpian family; the assertion will then follow from 9.2.4 and 9.4.1. Let f be an ω_1-isomorphism. For any element a' in the base of R′ and any denumerable ordinal α, there exists an element a in the base of R such that $f \cup \{(a, a')\}$ is an α-isomorphism. Since the set of possible choices for a is denumerably infinite, at least one of them satisfies this condition for arbitrarily large denumerable ordinals, hence for any denumerable ordinal. Thus the mapping is an ω_1-isomorphism.

Conversely, let a be a given element of the base R. For any denumerable ordinal α, there exists an element $a'(\alpha) \in |R'|$ such that $f \cup \{(a, a'(\alpha))\}$ is an α-isomorphism. By the preceding reasoning, there exists an element $b(\alpha) \in |R|$ such that $f \cup \{(b(\alpha), a'(\alpha))\}$ is an ω_1-isomorphism of R onto R′. Since there are only denumerably many elements in $|R|$, there is an element b satisfying the above condition for arbitrarily large denumerable ordinals α. Let F denote the domain of f. The identity mapping on F, extending by mapping a onto b, is an α-automorphism of R for arbitrarily large α, hence also an ω_1-automorphism. Consider any one of the elements $a'(\alpha)$, say a'. The mapping $f \cup \{(a, a')\}$ is the composite of the above ω_1-automorphism and the mapping $f \cup \{(b, a')\}$. Thus it is an ω_1-isomorphism of R onto R′. Hence these ω_1-isomorphisms form a Karpian family and they are therefore ∞-isomorphisms.

Since any ω_1-automorphism of R is an ∞-automorphism, it is extendible to an automorphism of R by 9.4.1. ◁

Problems. (1) Let R be a denumerable multirelation. Does there exist a denumerable ordinal α such that any α-automorphism *with denumerable domain* is extendible to an automorphism of R? (If the domain is finite, this follows from 9.4.2.)

(2) Let R be a denumerable multirelation. Does there exist a denumerable ordinal α such that any α-isomorphism of R onto an arbitrary multirelation, *with denumerable domain*, is an ∞-isomorphism? (If the domain is finite, consider the automorphic rank of R plus 2, and use 9.3.2 and 9.3.3.)

9.4.4. *Let* R, R′ *be two ∞-equivalent multirelations. For any denumerable subsets* $D \subseteq |R|$ *and* $D' \subseteq |R'|$, *there exist denumerable supersets* $D^+ \supseteq D$ *and* $D'^+ \supseteq D'$ *such that* $R \mid D^+$ *and* $R' \mid D'^+$ *are isomorphic.*

▷ Arrange the elements of D and D′ in sequences a_i and a'_i $(i = 0, 1, 2, \ldots)$ and then reason as in 9.4.1, starting from the empty mapping, which is by assumption an ∞-isomorphism. ◁

The converse is false: the ordinals $R = \omega_1$ and $R' = \omega_1 + 1$ satisfy the condition but they are not 2-equivalent, for R′ has a maximal element a' and there is no a such that the mapping $a \to a'$ is a 1-isomorphism.

Two relations R, R′ may have the same denumerable restrictions (up to isomorphism) and yet not satisfy the condition of the proposition. *Example:* let R be the chain of rational numbers and $R'=R+1$, $D=|R|$ and $D'=|R'|$.

9.4.5. *Two nondenumerable relations may be ∞-equivalent though neither is isomorphic to a restriction of the other.*

▷ Let R be the chain of real numbers, Q the chain of rationals. Define $R'=\omega_1(Q)$ as the chain obtained when each element of ω_1 is replaced by an isomorphic image of Q. Since both R and R′ are dense chains with neither minimal nor maximal element, they are ∞-equivalent, by 9.2.4. But no restriction of R is isomorphic to ω_1; on the other hand, if there were a restriction of R′ isomorphic to R, we could find two elements in the base of R′ defining an interval equipollent to the continuum, and this is impossible. ◁

9.4.6. *If* R *is a relation of the cardinal of the continuum, an ∞-automorphism of* R *need not have an extension which is an automorphism of* R.

▷ Consider the chain $Q+R$, where Q and R are as before. The mapping taking an element of the initial interval Q onto an element of the final interval R is not extendible to an automorphism of $Q+R$. Nevertheless, since the local automorphisms with finite domains form a Karpian family, they are ∞-automorphisms. ◁

9.4.7. *Let α be an ordinal. Then any relation $(\alpha+2)$-equivalent to α is isomorphic to α.*

▷ By 9.3.1, the automorphic rank of α is $\leqslant\alpha$. Let R be a relation which is $(\alpha+2)$-equivalent to α. By 9.3.3, the rank of R is $\leqslant\alpha$, and by 9.3.2 the relations α and R are ∞-equivalent. They have the same finite denumerable restrictions (up to isomorphism) (see 9.4.4), and thus R is well-ordered. Hence R is isomorphic to an ordinal. Finally, R is isomorphic to α by 9.2.2. ◁

9.4.8. *For any ordinal α there exist two α-equivalent relations not having the same denumerable restrictions (up to isomorphism).* This is the case for the chains ω^α and $\omega^\alpha+\omega^-(\omega^\alpha)$ considered in 9.1.5: ω^- is a restriction of the second but not of the first.

There exist two ω_1-equivalent relations of the cardinal of the continuum *which do not have the same denumerable restrictions (up to isomorphism).* Take $\alpha=\omega_1$ in the preceding example, to obtain $\omega_1+\omega^-(\omega_1)$.

9.4.9. *Let* R, R′ *be two multirelations. Even if every denumerable restriction of* R *is isomorphic to a restriction of* R′, R *need not have an extension which is* ∞*-equivalent to* R′ (compare the logical extension theorem 4.4.3 of [HEN, 1953]).

▷ Let $R=\omega_1$ be the smallest nondenumerable ordinal, E its base. For each element $a\in E$, we consider the restriction R_a of R to the elements $\leqslant a$. Let R′ be a partial order with base E′, formed by chains R'_a, each associated with an element $a\in E$ and isomorphic to R_a, with elements of two distinct component chains defined to be incomparable (mod R′). On the one hand, any denumerable restriction of R is isomorphic to a restriction of R′. On the other hand, suppose there exists an extension R^+ of R which is ∞-equivalent to R′. To the minimal element a of R corresponds an element a' of E′ defining an ∞-isomorphism of R^+ onto R′. Let b' be an R′-maximal element following a'; to b' corresponds an R^+-maximal element b following a; b thus follows all the elements of $R=\omega_1$, and the mapping taking (a, b) onto (a', b') is an ∞-isomorphism. Now let c be the element of E which defines the initial interval isomorphic to the interval from a' to b'. By repeated application of 9.2.1, we thus obtain an infinite decreasing sequence of elements of E; this is a contradiction (communicated by D. Scott). ◁

The example may be sharpened, requiring that R and R′ have the same denumerable restrictions (up to isomorphism): let R′ be as before and replace R by the common extension of R′ and one component ω_1.

Problem. Do there exist two multirelations with the same denumerable restrictions (up to isomorphism), such that neither has an extension ∞-equivalent to the other?

9.5. α-EXTENSION AND α-INTERPRETABILITY

9.5.1. Let R be a multirelation. An extension R′ of R is an *α-extension* if, for any finite subset F of |R|, the identity mapping on F is an α-isomorphism of R onto R′. Setting $\alpha=1$ we get previous notion of *1-extension* (for any finite subset F′ of |R′|, there exists a local isomorphism of R′ onto R, with domain F′, whose restriction to $F'\cap|R|$ is the identity mapping).

Problem (suggested by the upward Löwenheim-Skolem-Mal'tsev theorem). Let R be a multirelation with infinite base E, E′ a superset of E such that E′ − E is (at least) equipollent to E. Does R possess an α-extension with base E′? The only positive information available at present concerns the case $\alpha = 1$ (see Volume 1, 3.2.5).

9.5.2. An extension R′ of R is an ∞*-extension* if, for any finite subset F of |R|, the identity mapping on F is an ∞-isomorphism of R onto R′.

If R *has a denumerable base and* R′ *is any multirelation* ∞*-equivalent to* R, *then* R′ *is isomorphic to an* ∞*-extension of* R.

▷ Using 9.2.1, one can successively adjoin all the elements of |R| to the domain of the ∞-isomorphism in question. ◁

9.5.3. Given two multirelations R and S with the same base and arbitrary finite arities, we say that S is α*-interpretable* in R if every α-automorphism of R is a local automorphism of S. Similarly, one defines ∞*-interpretability*. It can be shown that the condition need hold only for α-automorphisms defined on at most n elements, where n is the arity of S (or the maximal arity of the relations in S).

If the base is denumerable, any ω_1*-automorphism is extendible to an automorphism* (see 9.4.3). *Consequently,* S *is* ω_1*-interpretable in* R *if and only if every automorphism of* R *is an automorphism of* S. Problem 9.4.3(1) reappears here: can one replace ω_1 by a denumerable ordinal α depending only on R?

Anticipating what follows we note that S is α-interpretable in R if and only if there exists an infinite formula with finite quantifiers, of rank $\leqslant \alpha$, which yields S when its predicates are replaced by R (see 9.6 below). This proposition is equivalent to a result of [SCO, 1965].

9.6. INFINITE LOGICAL CALCULI AND THEIR RELATION TO LOCAL ISOMORPHISMS AND EQUIVALENCES

The notion of infinite calculi goes back (at least) to [TAR, 1958].

We extend the concept of a relation as follows. Let I be an arbitrary set of *indices*. The relation to be defined will be called *finitary* or *infinitary* according as I is finite or infinite. The relation will associate a truth value + or − with each sequence of base elements indexed by the elements of I.

We first consider a *finitary multirelation*, i.e., a finite sequence of finitary relations with the same base. The definitions of free operators (Volume 1, 4.3) and connections are retained unchanged. However, we now allow *infinite conjunctions* and *disjunctions*, which refer to infinite sequences of relations with the same base. The quantifiers $\forall$ and $\exists$ are extended to apply to an arbitrary (finite or infinite) set of indices. The *truth value* of a formula, as defined in Volume 1, 5.2, extends immediately and naturally to the infinite case. Note that the *operator* represented by an infinite formula may transform the original finitary multirelation into an infinitary relation. The construction just described involves three important cardinals and an ordinal, which we now proceed to define.

If the conjunctions and disjunctions are allowed to refer to infinite sets of formulas of cardinal $<\alpha$, we call α the *connective length*. From the relational standpoint, the symbols of our formulas represent, first, the operators of the usual calculus: identifiers, finitary rank-changers, connectors. Also permitted are the following constructions. Given a set I of cardinal $<\alpha$, we can construct a multirelation which is an I-sequence of relations R_i (these relations are assumed to be either given from the start or constructed at previous stages), indexed by the elements of I. Let J_i denote the arity of R_i and J the union of the sets J_i, $i \in I$. We can then construct the conjunction R of arity J, where $R(x) = +$ for all J-tuples x such that $R_i(x_i) = +$ for each $i \in I$ (x_i denotes the corresponding J_i-tuple obtained from x). The construction of the disjunction is analogous.

It is sufficient to consider connective lengths α which are regular infinite cardinals, i.e., the union of any set of cardinal $<\alpha$ whose elements are sets of cardinal $<\alpha$ is again a set of cardinal $<\alpha$.

If the construction permits J-ary relations for any set J of cardinal $<\beta$, the cardinal β is called the *arity*. An important special case is that of *finitary* infinite formulas, where the arity is ω; in other words, the union J of the sets J_i in the above construction is always finite. Since the basis for this construction consists of finitary relations, the arity is at most equal to the connective length.

If quantifiers with index sets of cardinal $<\gamma$ are permitted, the cardinal γ is called the *quantifier depth*. We may always assume that the quantifier depth is at most equal to the arity (hence also to the connective length), since the construction is based on finitary relations.

If the arity of a bound formula (without free indices) is finite or ω, the same is true of the quantifier depth. If the arity is $\alpha > \omega$, the quantifier depth is equal to the arity. This is because the only way to go from a relation to another relation of (strictly) smaller arity is via a quantifier (or a rank-changer, which is necessarily applied to the original finitary relations).

The *rank* of a formula is an ordinal ρ defined inductively as follows. The rank of a free formula is 0. A connection of formulas of ranks ρ_i gives a formula of rank $\rho = \max_i(\rho_i)$. The formula obtained by prefixing a quantifier $\forall$ (or $\exists$) with arbitrarily many indices to a formula of rank ρ is of rank $\rho+1$. It is clear that the rank is at most equal to the connective length (assumed to be infinite).

Given two multirelations R and R′, a logical isomorphism f of R onto R′ with domain F and range $F' = f(F)$ will be called an $(\alpha, \beta, \gamma, \rho)$*-isomorphism* if it satisfies the following condition. Let P be a formula of connective length $\leqslant \alpha$, arity $\leqslant \beta$, quantifier depth $\leqslant \gamma$ and rank $\leqslant \rho$, with predicates replaceable by R (hence also by R′, which is necessarily of the same arity). Then substitution of R for the predicates and of arbitrary elements $a, b, \ldots$ of F for the free indices yields the same truth value as substitution of R′ and of the elements $f(a), f(b), \ldots$ of F′:

$$P(R)(a, b, \ldots) = P(R')(f(a), f(b), \ldots). \tag{$*$}$$

F may be a bound formula; this is necessarily the case when F is empty, and then we say that the multirelations are $(\alpha, \beta, \gamma, \rho)$*-equivalent.*

Note the different treatment accorded to the rank, on the one hand, and to the cardinals α, β, γ, on the other. Indeed, α is the smallest cardinal *strictly* greater than the cardinals of the conjunctions and disjunctions allowed in the calculus, β is the smallest cardinal *strictly* greater than the cardinals of the arities of the various relations, and γ is the smallest cardinal *strictly* greater than the cardinals of the index sets of quantifiers. By contrast, ρ is the smallest ordinal greater than *or equal to* the ranks of the formulas P. It would be useless to demand here that ρ be strictly greater than these ranks, for then, in the case of a limit ordinal ρ, whenever R, $a, b, \ldots$ and R′, $f(a), f(b), \ldots$ give the same value to all formulas for α, β, γ and all ordinals $<\rho$, they would give the same value for α, β, γ and ρ.

It is evident that $(\alpha, \beta, \gamma, \rho)$-isomorphism is preserved under inversion, composition of mappings, restriction (with respect to either the domain or the range). An isomorphism of R (defined on the entire base) onto R′ (with range exhausting the base), or any of its restrictions, is an $(\alpha, \beta, \gamma, \rho)$-isomorphism for all α, β, γ, ρ. $(\alpha, \beta, \gamma, \rho)$-equivalence is reflexive, symmetric and transitive.

In what follows we shall replace α by ∞ whenever the isomorphism (equivalence) under consideration is an $(\alpha, \beta, \gamma, \rho)$-isomorphism (-equivalence) for any cardinal α; similar conventions will be adopted for β, γ and the ordinal ρ.

9.6.1. The usual logical algebra of (k, p)-isomorphisms already calls for infinite formulas. Indeed, for a local isomorphism f of R onto R′, with *infinite domain* F, to be a (k, p)-isomorphism (where k and p are natural numbers), it is not sufficient that the restrictions of f to all finite subsets be (k, p)-isomorphisms. In other words, it is not sufficient that the equality $(*)$ should hold for any finite formula P and arbitrary elements $a, b, \ldots \in |R|$. For example, if f is an isomorphism of the (entire) chain of rationals Q onto a (proper) restriction of Q, it is clear that every restriction of f to a finite subset of the base is extendible to an automorphism of Q and is therefore a (k, p)-automorphism of Q for any k, p. However, f itself is not even a $(1, 1)$-automorphism of R.

To get a necessary and sufficient condition for f to be a (k, p)-isomorphism in this situation, we must require that equality $(*)$ hold for all formulas P with infinite conjunctions and disjunctions, of characteristic at most (k, p) (see 1.3 for the definition of characteristic). The proof that (k, p)-isomorphism is indeed characterized by this condition is analogous to the proof for a finite domain (see 1.3).

9.6.2. *Let α be an ordinal. Then α-isomorphism is identical with $(\infty, \infty, \omega, \alpha)$-isomorphism. α-equivalence is identical with $(\infty, \infty, \omega, \alpha)$-equivalence and $(\infty, \omega, \omega, \alpha)$-equivalence.*

▷ The assertion is obvious for $\alpha = 0$. Suppose that it is true for α and consider the case of $\alpha+1$. Let f be an $(\alpha+1)$-isomorphism of R onto R′ and consider a formula of rank $\alpha+1$, of the form $\forall_I P$, where P is of rank α (with finite quantifiers). Replace the predicates by R and substitute elements $a, \ldots$ of the domain of f for the free indices. Suppose that $\forall_I P(R)(a, \ldots) = +$. Then, for any finite sequence of elements $b', \ldots$ of the base of R′, indexed by elements of I, there exists a sequence of elements

$b, \ldots$ of $|\mathrm{R}|$, such that the mapping taking each a onto $f(a)$ and each b onto b' is an α-isomorphism, so that

$$\mathrm{P}(\mathrm{R}')(f(a), \ldots, b', \ldots) = \mathrm{P}(\mathrm{R})(a, \ldots, b, \ldots) = +.$$

Varying the elements b', we see that $\forall_{\mathrm{I}}\mathrm{P}(\mathrm{R}')(f(a), \ldots) = +$. The same argument is valid for the value $-$ and with R and R′ interchanged. It remains to observe that any formula of rank $\alpha+1$ is obtained from formulas of type $\forall_{\mathrm{I}}\mathrm{P}$, where P is of rank α, and from formulas of ranks $\leqslant\alpha$, by connections; thus they take the same value for $(\mathrm{R}; a, \ldots)$ and $(\mathrm{R}'; f(a), \ldots)$.

Conversely, let R, R′ and f be given and suppose that for all elements a in the domain the systems $(\mathrm{R}; a, \ldots)$ and $(\mathrm{R}'; f(a), \ldots)$ give the same value to any formula of rank $\alpha+1$ with finite quantifiers. Then, for any finite sequence of elements $b \in |\mathrm{R}|$ indexed by elements of I, and any formula P of rank α with finite quantifiers, the formula $\exists_{\mathrm{I}}\mathrm{P}$ assumes the same value for both systems. Let P be the conjunction of all formulas of ranks $\leqslant\alpha$ with finite quantifiers such that $\mathrm{P}(\mathrm{R})(a, \ldots, b, \ldots) = +$; for any formula of rank $\leqslant\alpha$ with finite quantifiers and of the same arity, either the formula itself or its negation must appear in this conjunction. By hypothesis, $\exists_{\mathrm{I}}\mathrm{P}(\mathrm{R}')(f(a), \ldots) = +$, and so there exists a sequence of elements b' such that $\mathrm{P}(\mathrm{R}')(f(a), \ldots, b', \ldots) = +$. Thus, $(\mathrm{R}; a, \ldots, b, \ldots)$ and $(\mathrm{R}'; f(a), \ldots, b', \ldots)$ give the same value to any formula of rank $\leqslant\alpha$ with finite quantifiers; by hypothesis, the mapping taking each a onto $f(a)$ and each b onto the corresponding b' is an α-isomorphism of R onto R′. Interchanging R and R′, we see that f is an $(\alpha+1)$-isomorphism.

Now let α be a limit ordinal. A local isomorphism f is an α-isomorphism if it is a β-isomorphism for every $\beta<\alpha$. On the other hand, systems $(\mathrm{R}; a, \ldots)$ and $(\mathrm{R}'; f(a), \ldots)$ give the same value to any formula of rank α with finite quantifiers if and only if they do so for any formula of rank $<\alpha$ with finite quantifiers. This completes the proof for $(\infty, \infty, \omega, \alpha)$-isomorphism and $(\infty, \infty, \omega, \alpha)$-equivalence.

The proof for $(\infty, \omega, \omega, \alpha)$-equivalence is similar, except that attention is now confined to formulas with finite quantifiers *and* finite arities, which are the only possible subformulas of a bound formula with finite quantifiers. $\lhd$

9.6.3. *Let* R *be a denumerable multirelation, a a finite sequence of elements of its base and* α *a denumerable ordinal. There exists a finitary formula* P *with denumerable conjunctions and disjunctions, with finite quantifiers, and of rank* α, *such that* $P(R')(a') = +$ *if and only if the mapping taking the sequence* a *onto* a' *is an* α*-isomorphism of* R *onto* R′.

Let R *and* α *be as before. Then there exists a bound formula* P *as described above, which is true only for multirelations* α*-equivalent to* R.

▷ The assertion is obvious for $\alpha=0$. Supposing it true for α, we now prove it for $\alpha+1$. If the mapping $a \to a'$ is an $(\alpha+1)$-isomorphism of R onto R′, this means that for any finite sequence b there is a sequence b' such that the mapping $a, b \to a', b'$ is an α-isomorphism of R onto R′, and conversely with R, a and R′, a' interchanged. With each finite set I of indices and each I-sequence b of elements of $|R|$, we associate a formula P_b of rank $\leqslant \alpha$ which satisfies the assertion for R, a, b. Since the set of sequences b is denumerable, the same is true of the set of formulas $\exists_I P_b$ (I may be a set of natural numbers, for example). For each finite set I, let Q_I denote the disjunction of the formulas P_b for all I-sequences b of elements of $|R|$, and consider the formula $\forall_I Q_I$, which is of rank $\leqslant \alpha+1$. Since the set of all possible sets I is denumerable, it is meaningful to consider the conjunction of all formulas $\exists_I P_b$ and $\forall_I Q_I$. If $(R'; a')$ satisfies this conjunction, the mapping taking a onto a' is an $(\alpha+1)$-isomorphism of R onto R′. Conversely, if this mapping is an $(\alpha+1)$-isomorphism, the conjunction just considered takes the same value for $(R; a)$ and $(R'; a')$ (see 9.6.2).

Now let α be a limit ordinal. If the assertion is true for R, a and every ordinal $\beta \leqslant \alpha$, let P_β denote the corresponding formulas. The (denumerable) conjunction of these formulas is true for $(R'; a')$ if and only if the mapping $a \to a'$ is a β-isomorphism for each $\beta < \alpha$, hence if and only if it is an α-isomorphism. ◁

9.6.4. *Let* R *be a denumerable multirelation of automorphic rank* α *(which is denumerable by* 9.3*). There exists a finitary formula* P, *with denumerable conjunctions and disjunctions, finite quantifiers, and of rank* $\alpha+2$, *such that any multirelation satisfying* P *is* ∞*-equivalent to* R, *and is therefore an* ∞*-extension of* R *up to isomorphism* (see [CAL, 1969]). *Moreover, any denumerable multirelation satisfying* P *is isomorphic to* R (see [SCO, 1965]).

▷ Consider the formula P whose existence was proved in 9.6.3, with

α replaced by $\alpha+2$. Then any multirelation R′ satisfying P is $(\alpha+2)$-equivalent to R, and therefore ∞-equivalent to R by 9.3.3 and 9.3.2. It is also an ∞-extension of R (up to isomorphism), by 9.5.2. Moreover, if R′ is denumerable, it is isomorphic to R by 9.4.1. $\lhd$

9.6.5. In conclusion, we wish to point out an important difference between the calculus of $(\infty, \omega, \omega, \rho<\omega_1)$-formulas and the calculus of $(\omega_1, \omega, \omega, \rho<\omega_1)$-formulas. For the first calculus, the example of 9.1.2 shows that Löwenheim's theorem fails: there exists an $(\omega_2, \omega, \omega, 2)$-formula which is true for the birelation (Z, C), which has the cardinal of the continuum, but false for every denumerable birelation. The same example shows that for the second calculus Skolem's theorem fails, except that one must replace the single formula in question by a continuum of formulas with denumerable conjunctions and disjunctions. However, Löwenheim's theorem is valid, as is the statement [SCO, 1965]: any formula of the second calculus having a model of infinite cardinal has a model of any smaller infinite cardinal. In fact, although the formula in question need not have a prenex form, we may assume that the negations govern only atomic subformulas, and then replace each existential quantifier by a Skolem function whose arguments are the indices of the quantifiers that dominate it and also the free indices (both these sets of indices are finite). Starting from an infinite subset D of the base and closing it with respect to Skolem functions, we obtain a superset equipollent to D.

9.6.6. The concept of (k, p)-isomorphism may be generalized to infinite ordinals k and p, as follows. A local isomorphism f of R onto R′ is a *$(0, p)$-isomorphism* for any ordinal p. If $k \geqslant 1$, f is a *(k, p)-isomorphism* if the following condition holds: for any ordinal $k'<k$, any finite sequence of ordinals $p'=p_1<p_2<\cdots<p_h<p$, and any sequence of elements $a_1, \ldots, a_h$ in $|R|$, there exists a sequence $a'_1, \ldots, a'_h \in |R'|$ such that the union of f and the pairs (a_i, a'_i) $(i=1, \ldots, h)$ is a (k', p')-isomorphism of R onto R′; and conversely.

Note that for any integer k a (k, ω)-isomorphism is a (k, p)-isomorphism for all $p<\omega$. For any ordinal α, an α-isomorphism as defined in 9.1 is an $(\alpha, \alpha(\omega))$-isomorphism, where $\alpha(\omega)$ is an α-sequence of ordinals isomorphic to ω.

APPENDIX

Proof of Lemmas Needed to Prove J. Robinson' Theorem [J. ROB, 1949]

Assuming only the classical theorems of arithmetic stated below in Sections 7 and 9, we shall prove Lemmas (1) and (2) of Volume 1, 7.3.3 (the lemmas are restated in Sections 12 and 13 below).

1. *Let n be a natural number $\geqslant 3$. An odd natural number is a quadratic residue of 2^n (i.e., a square $\bmod 2^n$) if and only if it is congruent to $+1 \bmod 8$.*

▷ Any odd square may be expressed as

$$(2a+1)^2 = 4a(a+1)+1,$$

where a is a natural number; thus it is congruent to $+1 \bmod 8$, and the same is true if we consider squares $\bmod 2^n$ ($n \geqslant 3$). Conversely, the number of integers from 0 to 2^n congruent to $+1 \bmod 8$ is 2^{n-3}; it will therefore suffice to show that this is the number of quadratic residues of 2^n. Indeed, if $0 \leqslant a < b < 2^{n-3}$ we have $(2a+1)^2 \not\equiv (2b+1)^2 \pmod{2^n}$, since $b-a$ and $a+b+1$ cannot both be even and neither of them is a multiple of 2^{n-2}. ◁

2. *Let n be a positive integer, p an odd prime. If $a \not\equiv 0 \pmod p$, then a is a quadratic residue of p^n if and only if it is a quadratic residue of p.*

▷ If $a \equiv b^2 \pmod{p^n}$, then $a \equiv b^2 \pmod p$. Conversely, consider the quadratic residues of p not divisible by p lying between 0 and p^n. Since the number of residues in any interval of length p is $\frac{1}{2}(p-1)$, the number of residues from 0 to p^n is $\frac{1}{2}(p-1)p^{n-1}$. It will now suffice to show that this is also the number of quadratic residues of p^n which are not divisible by p. Indeed, let a, b be such that $a, b \not\equiv 0 \pmod p$ and $0 < a < b < p^n/2$. The number of all such integers is $\frac{1}{2}(p-1)p^{n-1}$, and $a^2 \not\equiv b^2 \pmod{p^n}$, for $b-a$ and $a+b$ cannot both be multiples of p and neither of them is a multiple of p^n. ◁

3. Let p be a prime. A *p-adic integer* is a sequence a of integers a_n ($n=$

$=1, 2, \ldots$), called *components*, such that $0 \leqslant a_n < p^n$ and $a_{n+1} \equiv a_n \pmod{p^n}$. Addition and multiplication of p-adic integers is defined in componentwise fashion: $a_n + b_n \pmod{p^n}$ and $a_n b_n \pmod{p^n}$. We thus obtain an integral domain (commutative ring without divisors of zero). Each integer a is identified with the sequence whose components are $a_n \equiv a \pmod{p^n}$. A p-adic integer a is *invertible* (i.e., has an *inverse* b such that $ab = 1$) if and only if its first component a_1 is nonzero.

The field of quotients of the ring of p-adic integers is known as the *p-adic field* Q_p. Each rational number is identified with an element of Q_p by virtue of the above identification of its numerator and denominator with p-adic integers.

Each nonzero element of Q_p *is uniquely expressible as the product of* p^h *(where h is an integer) and an invertible p-adic integer.*

4. *Let u be the numerator, v the denominator of a rational number u/v. Then u/v is identified with an invertible p-adic integer if and only if neither u nor v is divisible by p.*

▷ The first component t_1 of $t = u/v$ is defined by $u \equiv vt_1 \pmod{p}$. This congruence has a solution t_1 if and only if $v \not\equiv 0 \pmod{p}$, and the solution is nonzero $\pmod{p}$ if and only if $u \not\equiv 0 \pmod{p}$. ◁

5. *A nonzero element of* Q_2 *is a square if and only if it is the product of* 2^h *(where h is an even integer) and a 2-adic integer all of whose components are congruent to* 1 (mod 8) *(this 2-adic integer is thus invertible).*

▷ By (3), any nonzero square in Q_2 is the product of 2^h (where h is even) and the square of an invertible 2-adic integer. The first component of the latter is 1, and so its first three components are (1, 1, 1), (1, 1, 5), (1, 3, 3) or (1, 3, 7). In these four cases the first components of the square are indeed (1, 1, 1). Conversely, by (1), any integer congruent to 1 (mod 8) is a square $\pmod{2^n}$ for any positive n. Finally, if the n-th component a and the $(n+1)$-th component b are odd numbers such that $a^2 \equiv b^2 \pmod{2^n}$, then either $b \equiv a \pmod{2^n}$, $-b \equiv a \pmod{2^n}$, or $\pm b \equiv a + 2^{n-1} \pmod{2^n}$, so that we may identify a and b. ◁

6. *Let p be an odd prime. A nonzero element of* Q_p *is a square if and only if it is the product of* p^h *(where h is an even integer) and a p-adic integer whose*

first component (and therefore each component) is a square (mod p) *but not divisible by p (this p-adic integer is thus invertible).*

$\triangleright$ By (3), a nonzero square in Q_p is the product of p^h (h even) and an invertible p-adic integer. The components of the square are therefore squares (mod p) and not multiples of p. Conversely, consider a p-adic integer whose first component is a nonzero square (mod p). Then each component is a square (mod p) and not a multiple of p, so that it is a square (mod p^n) for each positive integer n (see (2)). Finally, if a is the n-th component and b the $(n+1)$-th, then $a^2 \equiv b^2 \pmod{p^n}$, so that either $b \equiv a \pmod{p^n}$ or $-b \equiv a \pmod{p^n}$, and we may choose either b or $-b$. $\triangleleft$

7. Let $ax^2+by^2+cz^2+\cdots$ be a quadratic form with rational coefficients $a, b, c, \ldots$. The classical theorem of Hasse-Minkowski is as follows.

A nonzero rational number t is representable by the above quadratic form with rational x, y, z if and only if, for every prime r, the number t is representable by the same form with x, y, z,… elements of the r-adic field, and moreover t is so representable with x, y, z,… real numbers.

8. Given a prime r and two nonzero integers a, b, we define the Hilbert symbol $(a, b)_r$ to be 1 if $ax^2+by^2-z^2=0$ has a rational solution other than (0, 0, 0), and -1 otherwise. Set $a=r^\alpha u$ and $b=r^\beta v$ where α, β are natural numbers and u, v integers not divisible by r. We now set $\dfrac{u}{r}=1$ or -1 according as u is a quadratic residue of r or not. If r is an odd prime, we set $\varepsilon(r)=0$ or 1 according as r is congruent to 1 or 3 (mod 4). We assume the classical formula for the Hilbert symbol:

$$(a, b)_r=\left(\frac{u}{r}\right)^\beta\left(\frac{v}{r}\right)^\alpha(-1)^{\alpha\beta\varepsilon(r)}.$$

When $r=2$, we set $a=2^\alpha u$ and $b=2^\beta v$, where α, β are natural numbers and u, v odd integers. Set $\omega(u)=0$ *if* $u\equiv\pm1 \pmod 8$, $\omega(u)=1$ if $u\equiv\pm3 \pmod 8$. We assume the following classical formula:

$$(a, b)_2=(-1)^{\varepsilon(u)\varepsilon(v)+\alpha\omega(v)+\beta\omega(u)}.$$

9, We now state the following classical theorem on ternary quadratic forms:

A quadratic form $ax^2+by^2+cz^2$ represents a rational number t in the r-adic field if and only if either

$$(-1, -abc)_r=(a, b)_r\,(b, c)_r\,(a, c)_r$$

or t is not the product of $-abc$ and an r-adic square.

This theorem implies the following two propositions.

Let p be a prime, $p\equiv -1 \pmod 4$, and r an arbitrary prime. A nonzero rational number t is representable by $x^2+y^2-pz^2$, with x, y, z in Q_r, if and only if $r\neq 2$ and $r\neq p$, or else t is not the product of p and a square in Q_r ($r=2$ or $r=p$).

▷ In this special case the condition of the general theorem becomes $(-1, p)_r=(1, 1)_r$; in other words, $(-1, p)_r=1$, since the equation $x^2+y^2-z^2=0$ has the solution (0, 1, 1). If $r\neq 2$ and $r\neq p$, we have $\alpha=\beta=0$ and the formula of (8) gives the result 1.

If $r=p$, we have $\alpha=0$, $\beta=1$, $u=-1$; now -1 is not a quadratic residue of p, since $p\equiv -1 \pmod 4$; thus the result is -1. If $r=2$, we have $\alpha=\beta=0$, $u=-1$, $v=p\equiv -1 \pmod 4$, so that $\varepsilon(u)=\varepsilon(v)=1$; thus the result is -1. ◁

Let p be a prime, $p\equiv 1 \pmod 4$, q an odd prime which is not a quadratic residue of p, and r an arbitrary prime. A nonzero rational number t is representable by $x^2+qy^2-pz^2$, with x, y, z in Q_r, if and only if $r\neq p$ and $r\neq q$, or else t is not the product of pq and a square of Q_r ($r=p$ or $r=q$).

▷ In this case the general condition becomes

$$(-1, pq)_r=(1, q)_r\,(1, -p)_r\,(q, -p)_r,$$

Now $(1, a)_r=1$ for all r and a, since the equation $x^2+ay^2-z^2$ has the solution (1, 0, 1); thus we have $(-1, pq)_r=(q, -p)_r$. It can be shown via the formulas of (8) that $(-1, pq)_r=(-1, p)_r\,(-1, q)_r$; similarly,

$$(q, -p)_r=(q, -1)_r\,(q, p)_r=(-1, q)_r\,(-1, p)_r\,(-q, p)_r.$$

Thus we get $(p, -q)_r=1$. Suppose that $r\neq 2$, $r\neq p$ and $r\neq q$; then $\alpha=\beta=0$ and the result is 1. If $r=2$, then $\alpha=\beta=0$ and $\varepsilon(p)=0$, so that the result is 1.

Now let $r=p$. Then $\alpha=1$, $\beta=0$, and $v=-q=(-1)\,q$, which is not a quadratic residue of p since -1 is a quadratic residue; thus the result is -1. If $r=q$, we have $\alpha=0$, $\beta=1$, and $u=p$, which is not a quadratic residue of q; indeed, q is not a quadratic residue of p and the exponent $\frac{1}{4}(p-1)(q-1)$ figuring in the quadratic reciprocity theorem is even; thus the final result is -1. ◁

10. *Let p be a prime, $p \equiv -1 \pmod 4$. A nonzero rational number t is representable by $x^2+y^2-pz^2$, with x, y, z rational numbers, with the following exceptions: $t=ks^2$, where $k \equiv p \pmod 8$ and s is rational, or $t=pks^2$, where k is a quadratic residue of p and not divisible by p.*

▷ By the Hasse-Minkowski theorem (7), t is representable as required if and only if it is so representable in each field Q_r where r is a prime. By (9), this is true if $r \neq 2$ and $r \neq p$. It is true for $r=2$ if and only if t is not the product of p and a square in Q_2, and for $r=p$ if and only if t is not the product of p and a square in Q_p.

Consider the first exception. If t is the product of p and a nonzero square of Q_2, it follows from (5) that $t=p \cdot 2^h(u/v)$, where h is even, u and v are the reduced numerator and denominator of a rational number which is a 2-adic integer with components congruent to 1 (mod 8). This 2-adic integer is invertible. By (4), u and v are odd; hence $v^2 \equiv 1 \pmod 8$ and $uv=(u/v)\,v^2 \equiv 1 \pmod 8$. Set $k=puv \equiv p \pmod 8$ and $s=2^{h/2}/v$; then $t=ks^2$. Conversely, suppose that $t=ks^2$, where $k \equiv p \pmod 8$. The 2-adic expansion of p begins with (1, 3), and so p is invertible. Each component of its inverse satisfies the relations $kp' \equiv pp' \equiv 1 \pmod 8$. Thus kp' is a 2-adic integer, both invertible and a square (see (5)). Finally, t is the product of p and the q-adic square $kp's^2$.

We now consider the second exceptional case. If t is the product of p and a nonzero square of Q_p then, by (6), we have $t=p \cdot p^h(u/v)$, where h is even and u/v is an invertible p-adic integer whose components are quadratic residues of p not divisible by p. By (4), u and v are not divisible by p. The product $uv=(u/v)\,v^2$ is a quadratic residue of p and not divisible by p. Setting $k=uv$ and $s=p^{h/2}/v$, we get $t=pks^2$. Conversely, suppose that $r=pks^2$, where k is a quadratic residue of p and not a multiple of p. Then k and s^2 are both p-adic squares and the same is true of their product. ◁

11. *Let p be a prime, $p \equiv 1 \pmod 4$, q an odd prime which is not a quadratic residue of p. A nonzero rational number t is representable by $x^2+qy^2-pz^2$, with rational numbers x, y, z, with the following exceptions: $t=pks^2$, where k is a quadratic nonresidue of p and s is rational, or $t=qks^2$, where k is a quadratic nonresidue of q.*

▷ The rational number t is representable as required if and only if it is so representable in every field Q_r. By (9), this is true if $r \neq p$ and $r \neq q$. It is also true for $r=p$ if and only if t is not the product of pq and a square in Q_p, and for $r=q$ if and only if t is not the product of pq and a square in Q_q.

Consider the first exceptional case. If t is the product of pq and a nonzero square of Q_p, then by (6) we have $t=pq\cdot p^h(u/v)$, where h is even and u/v is an invertible p-adic integer whose components are quadratic residues of p and not multiples of p. By (4), u and v are not multiples of p. The product $uv=(u/v)\,v^2$ is a quadratic residue of p and not a multiple of p. We set $k=quv$, which is a quadratic nonresidue of p (since q is), and $s=p^{h/2}/v$; then $t=pks^2$. Conversely, suppose that $t=pks^2$ where k is a quadratic nonresidue of p. The first component of q in Q_p is nonzero and so q is an invertible p-adic integer. Its inverse q', like q itself, is not a p-adic square; thus the components of q' are quadratic nonresidues of p, and so those of kq' are quadratic residues of p but not multiples of p. Finally, t is the product of pq and the p-adic square $(kq')s^2$.

The second exceptional case is treated in the same way: since q is a nonresidue of p, it follows from the reciprocity theorem that p is a nonresidue of q (the exponent $\frac{1}{4}(p-1)\,(q-1)$ is again even). ◁

12. *Let p be a prime, $p\equiv -1 \pmod 4$, and t a rational number. The equation*

$$pt^2+2=x^2+y^2-pz^2 \tag{$*$}$$

has a rational solution if and only if the reduced denominator of t is divisible neither by 2 nor by p.

▷ Let $t=u/v$, where the numerator and denominator are relatively prime. The form $x^2+y^2-pz^2$ represents pt^2+2 if and only if it represents pu^2+2v^2. Suppose that v is odd and not divisible by p. Then pu^2+2v^2 is not divisible by p and is therefore not of the form pks^2 (recall that k is not a multiple of p). Moreover, $v\equiv 1$ or $v\equiv 3 \pmod 4$, and so $v^2\equiv 1 \pmod 4$ and $u^2\equiv 0$ or $u^2\equiv 1 \pmod 4$. Thus $pu^2+2v^2\equiv 1$ or $2 \pmod 4$, and so this number cannot have the form ks^2, where $k\equiv p\equiv 3 \pmod 4$. By (10), equation ($*$) has a rational solution.

Now let v be even, say $v=2v'$. Then u is odd, $u=2u'+1$. Then

$$pu^2+2v^2=8v'^2+p(4u'(u'+1)+1)\equiv p \pmod 8$$

is not representable by the right-hand side of equation ($*$) (use (10)).

Now let v be a multiple of p, say $v=pv'$. Then

$$pu^2+2v^2=p(u^2+2pv'^2),$$

where u is not divisible by p, so that $u^2+2pv'^2$ is not divisible by p. Moreover, since the latter integer is a quadratic residue of p, it follows that pu^2+2v^2 is not representable by the right-hand side of equation (*) (use (10)). ◁

13. *Let p be a prime, $p \equiv 1 \pmod 4$, q an odd prime which is a quadratic nonresidue of p, and t is a rational number. Then the equation*

$$pqt^2+2=x^2+qy^2-pz^2 \qquad (**)$$

has a rational solution if and only if the reduced denominator of t is divisible neither by p nor by q.

▷ Set $t=u/v$, where u and v are relatively prime. The form $x^2+qy^2-pz^2$ represents pqt^2+2 if and only if it represents pqu^2+2v^2. Suppose that v is not a multiple of p or of q. Then the same is true of pqu^2+2v^2, which is therefore not of the form qks^2 (where k is a nonresidue of q). By (11), equation (**) has a rational solution.

Now let v be a multiple of p, say $v=pv'$. Then

$$pqu^2+2v^2=p(qu^2+2pv'^2),$$

where u is not divisible by p. Hence $qu^2+2pv'^2$ is also not divisible by p. Moreover, q is a nonresidue of p, and so qu^2 and $qu^2+2pv'^2$ are also nonresidues of p. By (11), our equation has no rational solution. Similar reasoning holds if v is a multiple of q, with p and q interchanged. Note that p is a nonresidue of q by the reciprocity theorem and the fact that $\frac{1}{4}(p-1)(q-1)$ is even. ◁

The above account has benefited from suggestions of A. Seigneur and J. L. Seignouret.

Closure of a Relation

We propose to extend to relations the closure procedure whereby the chain of real numbers is derived from the chain of rationals. A few relevant problems of logic will be stated.

Let R be a relation with base E, of arbitrary arity. Recall that an R-*interval* is a subset D of E such that every local automorphism of the restriction R | D is extended by the identity mapping on E − D to a local automorphism of R (see Volume 1, Chapter 4, Exercise 6). We define an

R-*filter* to be a set $\mathscr{F}$ of nonempty R-intervals satisfying the following conditions: (1) any R-interval containing an element of $\mathscr{F}$ is also an element of $\mathscr{F}$; (2) the intersection of two elements of $\mathscr{F}$ is an element of $\mathscr{F}$ (it is an R-interval; see *loc. cit.*). A maximal R-filter will be called an R-*ultrafilter*. Any R-filter may be extended to an R-ultrafilter. An R-ultrafilter is said to be *trivial* if it consists of all R-intervals containing a singleton. If the R-ultrafilter is nontrivial, the intersection of all its elements is empty (and therefore an R-interval), and each element is infinite; indeed, any finite element could be partitioned into finitely many singletons, which are all R-intervals, and one of them is an element of the ultrafilter.

For a given R-interval D and an R-ultrafilter $\mathscr{F}$, either D intersects each element of $\mathscr{F}$, and is therefore an element of $\mathscr{F}$, or there exists an element of $\mathscr{F}$ disjoint from D. Consequently, if $\mathscr{F}$ and $\mathscr{F}'$ are distinct R-ultrafilters, there exist elements $D \in \mathscr{F}$ and $D' \in \mathscr{F}'$ such that $D \cap D' = \emptyset$.

We now complete the base E by embedding it in the set E^* of all R-ultrafilters. This is done by identifying each trivial R-ultrafilter with the element of E generating it; $E^* - E$ is thus the set of nontrivial ultrafilters. With each R-interval D we associate the set $D^* \supseteq D$ obtained by adjoining to D all the R-ultrafilters of which D is a member. The sets D^* possess the finite intersection property (compactness): if the intersection of some set of sets D^* is empty, then the intersection of some finite subset is empty.

With each nontrivial R-ultrafilter $\mathscr{F}$ we associate a relation $R(\mathscr{F})$ with base $E(\mathscr{F})$ and the same arity as R. The bases $E(\mathscr{F})$ are assumed to be disjoint from E and from one another. The relations $R(\mathscr{F})$ are subjected to the following condition, which may always be satisfied: for any finite subset F of $E(\mathscr{F})$ and any element $D \in \mathscr{F}$, there exists an isomorphism of the restriction $R(\mathscr{F}) \mid F$ onto a restriction of $R \mid D$. If $E(\mathscr{F})$ is finite, the proof that this is possible follows from the fact that there are only finitely many relations of the given arity with base $E(\mathscr{F})$, and that since the sets D are infinite we can always find a bijective mapping of $E(\mathscr{F})$ into each set D. If $E(\mathscr{F})$ is infinite, we replace it in turn by each of its finite subsets and then apply the coherence lemma (Volume 1, 3.1.3). Once the relations $R(\mathscr{F})$ have been chosen (some of them may have empty bases), the *closure* R^+ of R is unambiguously defined as follows on the union E^+ of E and all the set $E(\mathscr{F})$. Let n be the arity of R and $x_1, \ldots, x_n$ elements of E^+. If some x_i is in E, we replace it by $x_i' = x_i$ and say that it is fixed. Now group all x_i lying in the same set $E(\mathscr{F})$ together and consider their images under a local isomorphism of $R(\mathscr{F})$ onto $R \mid D$, where $D \in \mathscr{F}$. Denote these images by x_i'; we stipulate that different members D of different ultrafilters $\mathscr{F}$ be pairwise disjoint and contain no fixed elements x_i. We then set $R^+(x_1, \ldots, x_n) = R(x_1', \ldots, x_n')$; this value is independent of the specific intervals D and isomorphisms chosen.

If R is the chain of natural numbers, there is only one nontrivial R-ultrafilter – the set of complements of finite sets. If R is the chain of rationals, the nontrivial R-ultrafilters correspond to cuts and therefore to irrational numbers and to the two cuts determined by each rational number. Finally, if R takes the value + only, we obtain all the ultrafilters (in the usual sense) on the base E.

Problems. (1) Given a relation R and a nontrivial R-ultrafilter $\mathscr{F}$, associate an empty relation with every other nontrivial R-ultrafilter. Does there exist a nonempty relation $R(\mathscr{F})$ whose closure is a logical extension of R?

(2) Let $\mathscr{F}$ and $\mathscr{G}$ be two nontrivial R-ultrafilters, $R(\mathscr{F})$ and $R(\mathscr{G})$ relations each of which yields a logical extension of R via closure (the relations associated with the other R-ultrafilters are empty). Is the extension obtained by considering both $R(\mathscr{F})$ and $R(\mathscr{G})$ a logical extension of R?

Consider a set of filters $\mathscr{F}$ on a base E. Each element of a filter $\mathscr{F}$ is a subset of E. Call this set of filters *compact* if, for any choice function f such that $f(\mathscr{F})$ is an element of $\mathscr{F}$ for each $\mathscr{F}$, there exist finitely many $\mathscr{F}$'s such that the union of the sets $f(\mathscr{F})$ is E. A multirelation R on E is said to be *compact* if the set of all R-ultrafilters is compact.

For any set E, the set of all ultrafilters on E is a compact set (communicated by M. Jean). Any chain is a compact relation.

Let R be a relation satisfying the following condition: for any finite union A of R-intervals, the complement $|R| - A$ is a finite union of R-intervals; then R is a compact relation. The converse is true provided there are only $\aleph_0$ ultrafilters; it is not known if this is always true.

REFERENCES

Ax, J. and S. Kochen

1965: 'Diophantine Problems over Local Fields', *Am. J. Math.* **87**, 605–648.

1966: 'Diophantine Problems over Local Fields: III. Decidable Fields', *Ann. Math.* **83**, 437–456.

Bell, J. L. and A. B. Slomson

1969: *Models and Ultraproducts*, North-Holland, Amsterdam.

Bénéjam, J.-P.

1970: 'Une classification des isomorphismes locaux, et son application à la logique', *Comptes Rendus* **271 (A)**, 529–532.

Beth, E. W.

1953: 'On Padoa's Method in the Theory of Definition', *Proc. Kon. Nederl. Akad. Wetens.* **A 56**, No. 4, and *Indag. Math.* **15**, No. 4, 330–339.

1954: 'Observations métamathématiques sur les structures simplement ordonnées', in *Applications scientifiques de la logique mathématique*, Actes du 2e. Colloque international de logique mathématique, Paris, 1952, pp. 29–35.

Calais, J.-P.

1969: 'Isomorphismes locaux generalisés', *Comptes Rendus* **268 (A)**, 761–764.

1969′: 'La méthode de Fraïssé dans les langages infinis', *Comptes Rendus* **268 (A)**, 785–788 and 845–848.

Chang, C. C.

1959: 'On Unions of Chains of Models', *Proc. Am. Math. Soc.* **10**, 120–127.

Cohen, P.

1963: 'The Independence of the Continuum Hypothesis', *Proc. Nat. Acad. Sci. U.S.A.* **50**, 1143–1148 and **51**, 105–110.

1966: *Set Theory and the Continuum Hypothesis*, Benjamin, New York.

Craig, W.

1960: 'Bases for First-Order Theories and Subtheories', *J. Symbolic Logic* **25**, 97–142.

Craig, W. and R. L. Vaught

1958: 'Finite Axiomatizability Using Additional Predicates', *J. Symbolic Logic* **23**, 289–308.

Ehrenfeucht, A.

1957: 'On Theories Categorical In Power', *Fund. Math.* **14**, 241–248.

1957′: 'Application of Games to Some Problems of Mathematical Logic', *Bull. Acad. Polon. Sci.* **5**, 35–37.

Engeler, E.

1959: 'Äquivalenzklassen von *n*-tuplen', *Z. Math. Logik* **5**, 340–345.

1959′: 'A Characterization of Theories with Isomorphic Denumerable Models', *Notices Am. Math. Soc.* **6**, 161.

Ershov, Yu. L.

1963: 'Decidability of Elementary Theories of Certain Classes of Abelian Groups', *Algebra i Logika* **1**, No. 6, 37–41 (Russian).

Feferman, S.
1955: 'Sum Operations on Relational Systems. Product Operations on Relational Systems', *Bull. Am. Math. Soc.* **61**, 172.
1957: 'Some Recent Works of Ehrenfeucht and Fraïssé', in *Summer Institute of Symbolic Logic* (mimeographed), pp. 201–209.
Fraïssé, R.
1966: 'Une généralisation de l'ultraproduit', *J. Symbolic Logic* **31**, 235–244.
Frayne, T. E., A. C. Morel and D. S. Scott
1962: 'Reduced Direct Products', *Fund. Math.* **51**, 195–228.
Glassmire, W.
1970: 'A Problem in Categoricity', *Notices Am. Math. Soc.* **17**, 295.
Gödel, K.
1930: 'Die Vollständigkeit der Axiome des logischen Funktionenkalküls', *Monatsh. Math. Phys.* **37**, 349–360.
1931: 'Über formal unentscheidbare Sätze der Principia Mathematica und verwandter Systeme', *Monatsh. Math. Phys.* **38**, 173–198.
1940: *The Consistency of the Axiom of Choice and of the Generalized Continuum Hypothesis with the Axioms of Set Theory*, Princeton University Press, Princeton, N. J.
Grzegorczyk, A.
1971: 'Universal Decision and Proof Procedure for Theories Categorical in Aleph Zero', in *Colloque sur la démonstration automatique*, Paris, 1968.
Henkin, L.
1949: 'The Completeness of the First Order Functional Calculus', *J. Symbolic Logic* **14**, 42–48.
1953: 'Some Interconnections between Modern Algebra and Mathematical Logic', *Trans. Am. Math. Soc.* **74**, 410–427.
1955: 'On a Theorem of Vaught', *J. Symbolic Logic* **20**, 92–93.
Henson, W.
1972: 'Countable Homogeneous Relational Structures and $\aleph_0$-Categorical Theories', *J. Symbolic Logic* **37**, 494–500.
Hessenberg, G.
1906: *Grundbegriffe der Mengenlehre*, Göttingen. (For an important discussion of this work, see Bachmann, H., *Transfinite Zahlen* (2. Aufl.), Springer, Berlin, 1967, pp. 107–112.)
Heyting, A.
1930: 'Die formalen Regeln der intuitionistischen Logik', *S.-B. Preuss. Akad. Wiss., Phys.-Math. Kl.*, 42–56.
Kargapolov, M. I.
1963: 'On the Elementary Theory of Abelian Groups', *Algebra i Logika* **1**, No. 6, 26–36 (Russian).
Karp, C.
1964: *Languages with Expressions of Infinite Length*, North-Holland, Amsterdam.
1965: 'Finite-Quantifier Equivalence', in *The Theory of Models* (Proc. 1963 Internat. Symp., Berkeley), North-Holland, Amsterdam, pp. 407–412.
Keisler, J.
1960: 'Theory of Models with Generalized Atomic Formulas', *J. Symbolic Logic* **25**, 1–26.
1961: 'Ultraproducts and Elementary Classes', *Proc. Kon. Nederl. Akad. Wetens.* **A64** and *Indag. Math.* **23**, 477–495.

1971: *Model Theory for Infinitary Logic*, North-Holland, Amsterdam.
Kleene, S. C.
1952: 'Finite Axiomatizability of Theories in the Predicate Calculus, Using Additional Predicate Symbols', *Memoirs Am. Math. Soc.*, No. 10, 27–68.
Kochen, S.
1958: 'Filtration Systems', *Notices Am. Math. Soc.* **5**, 605 and 671–672.
1961: 'Ultraproducts in the Theory of Models', *Ann. Math.* (Series 2) **74**, 221–261.
Läuchli, H. and J. Leonard
1966: 'On the Elementary Theory of Linear Order', *Fund. Math.* **59**, 109–116.
Leśniewski, S.
1927: 'O podstawach matematyki' (On the Foundations of Mathematics), *Przegląd Filozoficzny* **300**, 164, and four other papers in the same journal (1928–1931).
Lindström, P.
1964: 'On Model Completeness', *Theoria* **30**, 183–196.
Lopez-Escobar, E. G. K.
1964: *Infinitely Long Formulas with Countable Quantifier Degrees*, Thesis, Berkeley.
Łos, J.
1954: 'On the Categoricity in Power of Elementary Deductive Systems and some Related Problems', *Colloq. Math.* **3**, No. 1, 58–62.
1955: 'Quelques remarques, théorèmes et problèmes sur les classes définissables d'algèbres', in *Mathematical Interpretation of Formal Systems*, North-Holland, Amsterdam, pp. 98–113.
Łos, J. and R. Suszko
1957: 'On the Extending of Models (IV)', *Fund. Math.* **44**, 52–60.
Mal'tsev, A. I.
1936: 'Untersuchungen aus dem Gebiete der mathematischen Logik', *Rec. Math. (Mat. Sbornik), Nouv. Sér.* **1**, 323–336.
Marelz, W.
1973: 'Consistance d'une hypothèse de Fraïssé sur la définissabilité dans un language du second ordre', *Compt. Rend.* **A 276**, 1147–1150 and 1169–1172.
Montagüe, R. and L. R. Vaught
1959: 'A Note on Theories with Selectors', *Fund. Math.* **47**, 243–247.
Morley, M.
1962: 'Categoricity in Power', *Notices Am. Math. Soc.* **9**, 218. (Full version published in 1965: *Trans. Am. Math. Soc.* **114**, 514–538.)
Morley, M. and R. L. Vaught
1962: 'Homogeneous Universal Models', *Math. Scand.* **11**, 37–57.
Peano, G.
1894–1908: *Formulaire de mathématiques* (introduction and five volumes, edited by Peano and written by him in collaboration with seven other authors), Turin.
Presburger, M.
1929: 'Über die Vollständigkeit eines gewissen Systems der Arithmetik ganzer Zahlen, in welchem die Addition als einzige Operation hervortritt', in *Comptes Rendus 1er. Congr. Math. Pays Slaves*, Warsaw, pp. 92–101.
Ribeiro, H.
1961: 'On the Universal Completeness of Classes of Relational Systems', *Arch. Math. Log. Grund.* **5**, 90–95.
Robinson, A.
1956: *Complete Theories*, North-Holland Publishing Company, Amsterdam.

1963: *Introduction to Model Theory and to the Metamathematics of Algebra*, North-Holland, Amsterdam.

Robinson, J.

1949: 'Definability and Decision Problems in Arithmetic', *J. Symbolic Logic* **14**, 98–114.

Ryll-Nardzewski, C.

1959: 'On Theories Categorical in Power $\leqslant\aleph_0$', *Bull. Acad. Polon. Sci., Sér. Math. Astron. Phys.* **7**, 545–548.

Scott, D. S.

1965: 'Logic with Denumerably Long Formulas and Finite Strings of Quantifiers', in *The Theory of Models* (Proc. 1963 Internat. Symp., Berkeley), North-Holland, Amsterdam, pp. 329–341.

Shelah, S.

1971: 'Every Two Elementarily Equivalent Models Have Isomorphic Ultrapowers', *Israel J. Math.* **10**, 224–233.

Shoenfield, J. R.

1967: *Mathematical Logic*, Addison-Wesley, Reading, Mass.

Suppes, P.

1957: *Introduction to Logic*, Van Nostrand, Princeton, N. J.

1960: *Axiomatic Set Theory*, Van Nostrand, Princeton, N. J.

Svenonius, L.

1959: '$\aleph_0$-categoricity in First Order Predicate Calculus', *Theoria* (Lund) **25**, 82–94.

1959′: 'A Theorem on Permutations in Models', *Theoria* (Lund) **25**, 173–178.

1960: *Some Problems in Logical Model-Theory*, Library of Theoria, No. IV, CWK Gleerup, Lund; Ejnar Munksgaard, Copenhagen.

Tarski, A.

1934: *Fund. Math.* **23**, 161 (remark at end of paper by Skolem).

1936: 'Der Wahrheitsbegriff in den formalisierten Sprachen', *Studia Phil.* **1**, 261–404; English translation by H. J. Woodger in *Logic, Semantics, Metamathematics*, Clarendon Press, Oxford.

1940: *The Completeness of Elementary Algebra and Geometry*, Hermann, Paris; re-printed 1967 by Institut Blaise Pascal, Paris.

1951: *A Decision Method for Elementary Algebra and Geometry*, 2nd edition, University of California, Berkeley and Los Angeles.

1952: 'Some Notions and Methods on the Borderline of Algebra and Metamathematics', in *Proc. Internat. Congr. Math., Cambridge, Mass., 1950*, Vol. 1, pp. 705–720.

1954–1955: 'Contributions to the Theory of Models', *Proc. Kon. Nederl. Akad. Wetens.* **A57** *(Indag. Math.* **16***)*, 572–588 and *Proc. Kon. Nederl. Akad. Wetens.* **A58** *(Indag. Math.* **17***)*, 56–64. Summary in *Bull. Am. Math. Soc.* **59** (1953), 390–391.

1958: 'Remarks on Predicate Logic with Infinitely Long Expressions', *Colloq. Math.* **6**, 171–176.

Tarski, A., A. Mostowski and R. M. Robinson

1953: *Undecidable Theories*, North-Holland, Amsterdam.

Tarski, A. and R. L. Vaught

1957: 'Arithmetical Extensions of Relational Systems', *Compositio Math.* **13**, 81–102.

Vaught, R. L.

1961: 'Denumerable Models of Complete Theories', in *Infinitistic Methods (Proc. Symp. on Foundations of Math., Warsaw, 1959)*, Państwowe Wydawnictwo Naukowe, Warsaw, pp. 303–321.

1963: 'Models of Complete Theories', *Bull. Am. Math. Soc.* **69**(3), 299–313.

Zermelo, E.

1908: 'Untersuchungen über die Grundlagen der Mengenlehre', *Math. Ann.* **65**, 261–281.

INDEX

(The figures refer to chapter and section)

SYNTHESE LIBRARY

Monographs on Epistemology, Logic, Methodology,
Philosophy of Science, Sociology of Science and of Knowledge, and on the
Mathematical Methods of Social and Behavioral Sciences

Editors:

DONALD DAVIDSON (The Rockefeller University and Princeton University)
JAAKKO HINTIKKA (Academy of Finland and Stanford University)
GABRIËL NUCHELMANS (University of Leyden)
WESLEY C. SALMON (University of Arizona)

1. J. M. BOCHEŃSKI, *A Precis of Mathematical Logic.* 1959, X + 100 pp.
2. P. L. GUIRAUD, *Problèmes et méthodes de la statistique linguistique.* 1960, VI + 146 pp.
3. HANS FREUDENTHAL (ed.), *The Concept and the Role of the Model in Mathematics and Natural and Social Sciences, Proceedings of a Colloquium held at Utrecht, The Netherlands, January 1960.* 1961, VI + 194 pp.
4. EVERT W. BETH, *Formal Methods. An Introduction to Symbolic Logic and the Study of Effective Operations in Arithmetic and Logic.* 1962, XIV + 170 pp.
5. B. H. KAZEMIER and D. VUYSJE (eds.), *Logic and Language. Studies dedicated to Professor Rudolf Carnap on the Occasion of his Seventieth Birthday.* 1962, VI + 256 pp.
6. MARX W. WARTOFSKY (ed.), *Proceedings of the Boston Colloquium for the Philosophy of Science, 1961–1962,* Boston Studies in the Philosophy of Science (ed. by Robert S. Cohen and Marx W. Wartofsky), Volume I. 1973, VIII + 212 pp.
7. A. A. ZINOV'EV, *Philosophical Problems of Many-Valued Logic.* 1963, XIV + 155 pp.
8. GEORGES GURVITCH, *The Spectrum of Social Time.* 1964, XXVI + 152 pp.
9. PAUL LORENZEN, *Formal Logic.* 1965, VIII + 123 pp.
10. ROBERT S. COHEN and MARX W. WARTOFSKY (eds.), *In Honor of Philipp Frank,* Boston Studies in the Philosophy of Science (ed. by Robert S. Cohen and Marx W. Wartofsky), Volume II. 1965, XXXIV + 475 pp.
11. EVERT W. BETH, *Mathematical Thought. An Introduction to the Philosophy of Mathematics.* 1965, XII + 208 pp.
12. EVERT W. BETH and JEAN PIAGET, *Mathematical Epistemology and Psychology.* 1966, XII + 326 pp.
13. GUIDO KÜNG, *Ontology and the Logistic Analysis of Language. An Enquiry into the Contemporary Views on Universals.* 1967, XI + 210 pp.
14. ROBERT S. COHEN and MARX W. WARTOFSKY (eds.), *Proceedings of the Boston Colloquium for the Philosophy of Science 1964–1966, in Memory of Norwood Russell Hanson,* Boston Studies in the Philosophy of Science (ed. by Robert S. Cohen and Marx W. Wartofsky), Volume III. 1967, XLIX + 489 pp.
15. C. D. BROAD, *Induction, Probability, and Causation. Selected Papers.* 1968, XI + 296 pp.
16. GÜNTHER PATZIG, *Aristotle's Theory of the Syllogism. A Logical-Philosophical Study of Book A of the Prior Analytics.* 1968, XVII + 215 pp.
17. NICHOLAS RESCHER, *Topics in Philosophical Logic.* 1968, XIV + 347 pp.

18. Robert S. Cohen and Marx W. Wartofsky (eds.), *Proceedings of the Boston Colloquium for the Philosophy of Science 1966–1968*, Boston Studies in the Philosophy of Science (ed. by Robert S. Cohen and Marx W. Wartofsky), Volume IV. 1969, VIII + 537 pp.
19. Robert S. Cohen and Marx W. Wartofsky (eds.), *Proceedings of the Boston Colloquium for the Philosophy of Science 1966–1968*, Boston Studies in the Philosophy of Science (ed. by Robert S. Cohen and Marx W. Wartofsky), Volume V. 1969, VIII + 482 pp.
20. J. W. Davis, D. J. Hockney, and W. K. Wilson (eds.), *Philosophical Logic*. 1969, VIII + 277 pp.
21. D. Davidson and J. Hintikka (eds.), *Words and Objections: Essays on the Work of W. V. Quine*. 1969, VIII + 366 pp.
22. Patrick Suppes, *Studies in the Methodology and Foundations of Science. Selected. Papers from 1911 to 1969*, XII + 473 pp.
23. Jaakko Hintikka, *Models for Modalities. Selected Essays*. 1969, IX + 220 pp.
24. Nicholas Rescher *et al.* (eds.), *Essay in Honor of Carl G. Hempel. A Tribute on the Occasion of his Sixty-Fifth Birthday*. 1969, VII + 272 pp.
25. P. V. Tavanec (ed.), *Problems of the Logic of Scientific Knowledge*. 1969, XII + 429 pp.
26. Marshall Swain (ed.), *Induction, Acceptance, and Rational Belief*. 1970, VII + 232 pp.
27. Robert S. Cohen and Raymond J. Seeger (eds.), *Ernst Mach; Physicist and Philosopher*, Boston Studies in the Philosophy of Science (ed. by Robert S. Cohen and Marx W. Wartofsky), Volume VI. 1970, VIII + 295 pp.
28. Jaakko Hintikka and Patrick Suppes, *Information and Inference*. 1970, X + 336 pp.
29. Karel Lambert, *Philosophical Problems in Logic. Some Recent Developments*. 1970, VII + 176 pp.
30. Rolf A. Eberle, *Nominalistic Systems*. 1970, IX + 217 pp.
31. Paul Weingartner and Gerhard Zecha (eds.), *Induction, Physics, and Ethics, Proceedings and Discussions of the 1968 Salzburg Colloquium in the Philosophy of Science*. 1970, X + 382 pp.
32. Evert W. Beth, *Aspects of Modern Logic*. 1970, XI + 176 pp.
33. Risto Hilpinen (ed.), *Deontic Logic: Introductory and Systematic Readings*. 1971, VII + 182 pp.
34. Jean-Louis Krivine, *Introduction to Axiomatic Set Theory*. 1971, VII + 98 pp.
35. Joseph D. Sneed, *The Logical Structure of Mathematical Physics*. 1971, XV + 311 pp.
36. Carl R. Kordig, *The Justification of Scientific Change*. 1971, XIV + 119 pp.
37. Milič Čapek, *Bergson and Modern Physics*, Boston Studies in the Philosophy of Science (ed. by Robert S. Cohen and Marx W. Wartofsky), Volume VII. 1971, XV + 414 pp.
38. Norwood Russell Hanson, *What I do not Believe, and other Essays*, ed. by Stephen Toulmin and Harry Woolf. 1971, XII + 390 pp.
39. Roger C. Buck and Robert S. Cohen (eds.), *PSA 1970. In Memory of Rudolf Carnap*, Boston Studies in the Philosophy of Science (ed. by Robert S. Cohen and Marx W. Wartofsky), Volume VIII. 1971, LXVI + 615 pp. Also available as a paperback.
40. Donald Davidson and Gilbert Harman (eds.), *Semantics of Natural Language*. 1972, X + 769 pp. Also available as a paperback.
41. Yehosua Bar-Hillel (ed.), *Pragmatics of Natural Languages*. 1971, VII + 231 pp.
42. Sören Stenlund, *Combinators, λ-Terms and Proof Theory*. 1972, 184 pp.
43. Martin Strauss, *Modern Physics and Its Philosophy. Selected Papers in the Logic, History, and Philosophy of Science*. 1972, X + 297 pp.

44. MARIO BUNGE, *Method, Model and Matter*. 1973, VII + 196 pp.
45. MARIO BUNGE, *Philosophy of Physics*. 1973, IX + 248 pp.
46. A. A. ZINOV'EV, *Foundations of the Logical Theory of Scientific Knowledge* (*Complex Logic*), Boston Studies in the Philosophy of Science (ed. by Robert S. Cohen and Marx W. Wartofsky), Volume IX. Revised and enlarged English edition with an appendix, by G. A. Smirnov, E. A. Sidorenka, A. M. Fedina, and L. A. Bobrova 1973, XXII + 301 pp. Also available as a paperback.
47. LADISLAV TONDL, *Scientific Procedures*, Boston Studies in the Philosophy of Science (ed. by Robert S. Cohen and Marx W. Wartofsky), Volume X. 1973, XII + 268 pp. Also available as a paperback.
48. NORWOOD RUSSELL HANSON, *Constellations and Conjectures*, ed. by Willard C. Humphreys, Jr. 1973, X + 282 pp.
49. K. J. J. HINTIKKA, J. M. E. MORAVCSIK, and P. SUPPES (eds.), *Approaches to Natural Language. Proceedings of the 1970 Stanford Workshop on Grammar and Semantics*. 1973, VIII + 526 pp. Also available as a paperback.
50. MARIO BUNGE (ed.), *Exact Philosophy – Problems, Tools, and Goals*. 1973, X + 214 pp.
51. RADU J. BOGDAN and ILKKA NIINILUOTO (eds.), *Logic, Language, and Probability*. A selection of papers contributed to Sections IV, VI, and XI of the Fourth International Congress for Logic, Methodology, and Philosophy of Science, Bucharest, September 1971. 1973, X + 323 pp.
52. GLENN PEARCE and PATRICK MAYNARD (eds.), *Conceptual Chance*. 1973, XII + 282 pp.
53. ILKKA NIINILUOTO and RAIMO TUOMELA, *Theoretical Concepts and Hypothetico-Inductive Inference*. 1973, VII + 264 pp.
54. ROLAND FRAÏSSÉ, *Course of Mathematical Logic* – Volume I: *Relation and Logical Formula*. 1973, XVI + 186 pp. Also available as a paperback.
55. ADOLF GRÜNBAUM, *Philosophical Problems of Space and Time*. Second, enlarged edition, Boston Studies in the Philosophy of Science (ed. by Robert S. Cohen and Marx W. Wartofsky), Volume XII. 1973, XXIII + 884 pp. Also available as a paperback.
56. PATRICK SUPPES (ed.), *Space, Time, and Geometry*. 1973, XI + 424 pp.
57. HANS KELSEN, *Essays in Legal and Moral Philosophy*, selected and introduced by Ota Weinberger. 1973, XXVIII + 300 pp.
58. R. J. SEEGER and ROBERT S. COHEN (eds.), *Philosophical Foundations of Science. Proceedings of an AAAS Program, 1969*. Boston Studies in the Philosophy of Science (ed. by Robert S. Cohen and Marx W. Wartofsky), Volume XI. 1974, X + 545 pp. Also available as paperback.
59. ROBERT S. COHEN and MARX W. WARTOFSKY (eds.), *Logical and Epistemological Studies in Contemporary Physics*, Boston Studies in the Philosophy of Science (ed. by Robert S. Cohen and Marx W. Wartofsky), Volume XIII. 1973, VIII + 462 pp. Also available as paperback.
60. ROBERT S. COHEN and MARX W. WARTOFSKY (eds.), *Methodological and Historical Essays in the Natural and Social Sciences. Proceedings of the Boston Colloquium for the Philosophy of Science, 1969–1972*, Boston Studies in the Philosophy of Science (ed. by Robert S. Cohen and Marx W. Wartofsky), Volume XIV. 1974, VIII + 405 pp. Also available as paperback.
61. ROBERT S. COHEN, J. J. STACHEL and MARX W. WARTOFSKY (eds.), *For Dirk Struik. Scientific, Historical and Political Essays in Honor of Dirk J. Struik*, Boston Studies in the Philosophy of Science (ed. by Robert S. Cohen and Marx W. Wartofsky), Volume XV. 1974, XXVII + 652 pp. Also available as paperback.
62. KAZIMIERZ AJDUKIEWICZ, *Pragmatic Logic*, transl. from the Polish by Olgierd Wojtasiewicz.

63. SÖREN STENLUND (ed.), *Logical Theory and Semantic Analysis. Essays Dedicated to Stig Kanger on His Fiftieth Birthday*. 1974, V+217 pp.
64. KENNETH F. SCHAFFNER and ROBERT S. COHEN (eds.), *Proceedings of the 1972 Biennial Meeting, Philosophy of Science Association*, Boston Studies in the Philosophy of Science (ed. by Robert S. Cohen and Marx W. Wartofsky), Volume XX. 1974, IX+444 pp. Also available as paperback.
65. HENRY E. KYBURG, JR., *The Logical Foundations of Statistical Inference*. 1974, IX+421 pp.
66. MARJORIE GRENE, *The Understanding of Nature: Essays in the Philosophy of Biology*, Boston Studies in the Philosophy of Science (ed. by Robert S. Cohen and Marx W. Wartofsky), Volume XXIII. 1974, XII+360 pp. Also available as paperback.
67. JAN M. BROEKMAN, *Structuralism: Moscow, Prague, Paris*.
68. NORMAN GESCHWIND, *Selected Papers on Language and the Brain*, Boston Studies in the Philosophy of Science (ed. by Robert S. Cohen and Marx W. Wartofsky), Volume XVI. Also available as paperback.
69. ROLAND FRAÏSSÉ. *Course of Mathematical Logic* – Volume II: *Model Theory*. 1974, XIX+192 pp.
70. ANDRZEJ GRZEGORCZYK, *An Outline of Mathematical Logic*. Fundamental Results and Notions Explained with All Details. 1974, X+596 pp.

SYNTHESE HISTORICAL LIBRARY

Texts and Studies
in the History of Logic and Philosophy

Editors:

1. M. T. BEONIO-BROCCHIERI FUMAGALLI, *The Logic of Abelard.* Translated from the Italian. 1969, IX + 101 pp.

2. GOTTFRIED WILHELM LEIBNITZ, *Philosophical Papers and Letters.* A selection translated and edited, with an introduction, by Leroy E. Loemker. 1969, XII + 736 pp.

3. ERNST MALLY, *Logische Schriften,* ed. by Karl Wolf and Paul Weingartner. 1971, X + 340 pp.

4. LEWIS WHITE BECK (ed.), *Proceedings of the Third International Kant Congress.* 1972, XI + 718 pp.

5. BERNARD BOLZANO, *Theory of Science,* ed. by Jan Berg. 1973, XV + 398 pp.

6. J. M. E. MORAVCSIK (ed.), *Patterns in Plato's Thought. Papers arising out of the 1971 West Coast Greek Philosophy Conference.* 1973, VIII + 212 pp.

7. NABIL SHEHABY, *The Propositional Logic of Avicenna: A Translation from al-Shifā' :al-Qiyās,* with Introduction, Commentary and Glossary. 1973, XIII + 296 pp.

8. DESMOND PAUL HENRY, *Commentary on De Grammatico: The Historical-Logical Dimensions of a Dialogue of St. Anselm's.* 1974, IX + 345 pp.

9. JOHN CORCORAN, *Ancient Logic and Its Modern Interpretations.* 1974. X + 208 pp.

10. E. M. BARTH, *The Logic of the Articles in Traditional Philosophy.* 1974, XXVII + 533 pp.

11. JAAKKO HINTIKKA, *Knowledge and the Known. Historical Perspectives in Epistemology.* 1974, XII + 243 pp.

12. E. J. ASHWORTH, *Language and Logic in the Post-Medieval Period.* 1974, XIII + 304 pp.